I0889160

MATHEMATICAL RESEARCH | MATHEMATISCHE FORSCHUNG

Preprimal Algebras

K. Denecke

Band 11

THEORIA CUM PRAXI

Akademie-Verlag · Berlin

In this series original contributions of mathematical research in all fields are contained, such as

— research monographs
— collections of papers to a single topic
— reports on congresses of exceptional interest for mathematical research.

This series is aimed at promoting quick information and communication between mathematicians of the various special branches.

In diese Reihe werden Originalbeiträge zu allen Gebieten der mathematischen Forschung aufgenommen wie

— Forschungsmonographien
— Sammlungen von Arbeiten zu einem speziellen Thema
— Berichte von Tagungen, die für die mathematische Forschung besonders aktuell sind.

Die Reihe soll die schnelle Information und gute Kommunikation zwischen den Mathematikern der verschiedenen Fachgebiete fördern.

Manuscripts
in English and German comprising at least 100 pages and not more than 500 pages can be admitted to this series. With respect to a quick publication the manuscripts are reproduced photomechanically. Authors who are interested in this series please turn directly to the 'Akademie-Verlag'. Here you will get more detailed information about the form of the manuscripts and the modalities of publication.

Manuskripte
in englischer und deutscher Sprache, die mindestens 100 Seiten und nicht mehr als 500 Seiten umfassen, können in diese Reihe aufgenommen werden. Im Interesse einer schnellen Publikation werden die Manuskripte auf fotomechanischem Weg reproduziert. Autoren, die an der Veröffentlichung entsprechender Arbeiten in dieser Reihe interessiert sind, wenden sich bitte direkt an den Akademie-Verlag. Sie erhalten dort genauere Informationen über die Gestaltung der Manuskripte und die Modalitäten der Veröffentlichung.

K. Denecke

Preprimal Algebras

Mathematical Research   **Mathematische Forschung**

Wissenschaftliche Beiträge
herausgegeben von der
Akademie der Wissenschaften der DDR
Institut für Mathematik

Band 11

Preprimal Algebras

by
K. Denecke

Preprimal Algebras

by
Klaus Denecke

Akademie-Verlag Berlin 1982

Verfasser:
Dr. Klaus Denecke
Akademie der Landwirtschaftswissenschaften der DDR
Institut für Züchtungsforschung Quedlinburg
DDR-4300 Quedlinburg
E.-u.-J.-Rosenberg-Str. 22/23

Die Titel dieser Schriftenreihe werden vom Originalmanuskript der Autoren reproduziert.

Erschienen im Akademie-Verlag, DDR-1086 Berlin, Leipziger Str. 3—4
Lektor: Dr. Reinhard Höppner
© Akademie-Verlag Berlin 1982
Lizenznummer: 202 100/414/82
Gesamtherstellung: VEB Kongreß- und Werbedruck, 9273 Oberlungwitz
Umschlag: Rolf Kunze
Bestellnummer: 763 103 7 (2182/11) LSV 1025
Printed in GDR
DDR 22,— M

Hinweis an unsere Leser!

Wir freuen uns, mit der Monographie von Herrn Dr. Klaus Denecke über "Preprimal Algebras" wieder eine Publikation vorlegen zu können, die in der Reihe "Mathematical Research - Mathematische Forschung" speziell auf die Algebra und ihre Anwendungen eingeht.

In den Jahren 1976 bis 1979 war diesem Problemkreis eine besondere Reihe "Studien zur Algebra und ihre Anwendungen" gewidmet, in der die auf der folgenden Seite aufgeführten 7 Bände erschienen sind. Da diese Reihe ihr Erscheinen eingestellt hat, haben die Fachwissenschaftler nun die Möglichkeit, ihre Forschungsergebnisse auch auf dem Gebiet der Algebra in der vorliegenden Schriftenreihe zu veröffentlichen. In der Themenauswahl für dieses Teilgebiet beraten den Verlag weiterhin Herr Professor Dr. Lothar Budach und Herr Dr. Hans-Jürgen Hoehnke, denen wir für ihre Initiative danken.

Wir würden uns freuen, wenn die Fachspezialisten und Studierenden diesen Themen in der "Mathematischen Forschung" ihr Interesse zuwenden würden.

Akademie-Verlag

In der Reihe

STUDIEN ZUR ALGEBRA UND IHRE ANWENDUNGEN

sind in den Jahren 1976 bis 1979 folgende Bände erschienen:

Band 1
H.-J. Hoehnke (Hrsgb.):
Universelle Algebren und Theorie der Radikale
1976, 91 Seiten, 18,-- M, Best.-Nr. 762 223 7 (2174/1)

Band 2
G. A. Saizew
Algebraische Probleme der mathematischen und theoretischen Physik
1979, 164 Seiten, 32,-- M, Best.-Nr. 762 245 6 (2174/2)

Band 3
G. Richter
Kategorielle universelle Algebren
1979, 307 Seiten, 58,-- M, Best.-Nr. 762 377 4 (2174/3)

Band 4
J. A. Hahturin
Lectures on Lie Algebras
1978, 134 Seiten, 28,-- M, Best.-Nr. 762 492 9 (2174/4)

Band 5
W. Wechler
The Concept of Fuzziness in Automata and Language Theory
1978, 148 Seiten, 27,-- M, Best.-Nr. 762 517 1 (2174/5)

Band 6
H. Werner
Discriminator-Algebras
1978, 97 Seiten, 18,-- M, Best.-Nr. 762 519 8 (2174/6)

Band 7
H.-J. Hoehnke (Hrsgb.)
Algebraische Modelle, Kategorien und Gruppoide
1979, 178 Seiten, 36,-- M, Best.-Nr. 762 642 2 (2174/7)

Preface

The development of an algebraic structure theory for preprimal algebras appears as a problem of growing importance in connection with an algebraic approach to the solution of the functional completeness problem.

Completeness questions appeared first in the theory of two- and multiple-valued logics. This direction is characterized by the names of E.L. Post, S.W.Jablonskij, and I.G.Rosenberg. Completeness questions in universal algebras were first investigated by A.L.Foster, A.F.Pixley and R.W.Quackenbush. The recent attempts of a unified treatment of this subject were mainly supported by R.A.Knoebel.

The aim of this monograph is an extensive description of the actual state of the research into preprimal algebras on a unified base covering both dispositions.

My interest in preprimal algebras was stimulated by my tutor Dr.habil.H.-J.Hoehnke. Thanks to him for his helpful comments and permanent attention in my scientific development! My thanks are also directed to Dr.sc.R.Pöschel, who made corrections to a former version of this monograph.

Further I want to thank Prof.Dr.A.F.Pixley, Prof.Dr.H.Werner, Prof.Dr.I.G.Rosenberg, Dr.R.A.Knoebel, and Dr.R.W.Quackenbush. Their encouragement as well as their own scientific results have contributed to the accomplishment of this monograph.

Finally, I also want to acknowledge Dr.R.Höppner for friendly cooperation and including this publication in series „Mathematical Research".

Thale(Harz), March 1982 Klaus Denecke

Contents

Introduction 11

1. Basic concepts 26
1.1 Functions on a finite set 26
1.2 Finite algebras 30
1.3 Varieties 34
1.4 Categorical algebras 37
1.5 Paraprimal and cryptoprimal algebras 39
1.6 On finite axiomatizability of finite algebras 41

2. Arithmetical varieties 42
2.1 Characterization of quasiprimal and functionally complete algebras 42
2.2 A characterization of semiprimal, demiprimal, and primal algebras 46
2.3 Hemiprimal algebras 49
2.4 Representation theorem 51
Problems and comments 52

3. Two-element preprimal algebras 53
3.2 Two-element algebras and varieties generated by two-element algebras 56
3.3 Completeness properties of two-element algebras 63
Problems and comments 64

4. Preprimal algebras 65
4.1 Classification of all maximal classes of functions in $\underline{F}(A)$ 65
4.2 Elementary properties of preprimal algebras 69
4.3 Independent sets and free preprimal algebras 72
4.4 Weak automorphisms of preprimal algebras 75
Problems and comments 79

5. Varieties generated by preprimal algebras 80.
5.1 An algebraic characterization of preprimal algebras generating arithmetical varieties 80
5.2 Preprimal algebras generating either congruence distributive or congruence permutable varieties 83
5.3 Preprimal algebras generating neither congruence distributive nor congruence permutable varieties 89
5.4 Algebraic characterization of preprimal algebras generating 2-distributive varieties 91
5.5 Subdirectly irreducible elements, representation theorems 97

5.6 Equationally maximal preprimal algebras 108
5.7 Semidegenerate varieties generated by preprimal algebras 111
5.8 The lattice of subvarieties of a variety generated by a preprimal algebra 114
5.9 On finite axiomatizability of preprimal algebras 117
Problems and comments 118

6. Completeness in universal algebras 119
6.1 Primality 119
6.2 Primality and preprimality of finite algebras with a single fundamental operation 121
6.3 Functionally complete algebras 125
6.4 A new proof of Rosenberg's primal algebra characterization theorem by R.W.Quackenbush 130
Problems and comments 131

7. Examples for preprimal algebras 132

Appendix: Quackenbush's new proof of Rosenberg's theorem 137

Bibliography 153

Subject Index 161

Introduction

The purpose of this book is to enlarge the natural interconnections between algebra and logic. Concepts, methods, and results of Universal algebra are applied to describe and characterize preprimal algebras, a class of universal algebras which originates from multiple-valued logic. The investigations correspond to an international trend to consider finite algebra and multiple-valued logic in a close link. Moreover, they generalize a tradition going back to Boole and his followers to solve problems of classical logic using methods of the theory of Boolean algebras. To make the aim of this book understandable for the reader now we clarify some aspects of the interconnections between classical logic and Boolean algebras and also between multiple-valued logic and another class of algebras, the so-called Post algebras.

The propositional functions of the classical two-valued propositional calculus, the Boolean functions, are the functions on the two-element set $A = \{0,1\}$. Here the element 0 corresponds to the truth-value „false" and 1 corresponds to the truth-value „true". The „elementary" Boolean functions which correspond to conjunction, disjunction, implication, equivalence, and negation in the propositional calculus, are examples for Boolean functions. These Boolean functions are described by the following truth-values tables:

x_1	x_2	$x_1 \wedge x_2$	$x_1 \vee x_2$	$x_1 \supset x_2$	$x_1 \equiv x_2$
0	0	0	0	1	1
0	1	0	1	1	0
1	0	0	1	0	0
1	1	1	1	1	1

x	Nx
0	1
1	0

By the composition of Boolean functions, often also denoted as superposition, further Boolean functions can be derived. For instance, the following Boolean functions are derivable by superposition from the above defined elementary Boolean functions: $N(x_1 \wedge x_2)$, $((x_1 \vee x_2) \vee (x_2 \wedge x_3)) \supset (x_1 \vee x_3)$. It is natural to ask whether all Boolean functions can be derived from the elementary Boolean functions. More general: Characterize all subsets of the set of all Boolean functions such that from these subsets all Boolean functions can be derived by superposition. Such subsets are called functionally complete. Other questions are: Which one-element subsets $\{f\}$ of Boolean functions are functionally complete ? Let F be a set of Boolean functions. What is the set $\bar{F}$ of all Boolean functions which can be obtained as finitary superpositions of functions from F ? (These classes $\bar{F}$ are called closed under superposition.) Characterize all closed classes of Boolean functions. How large is the number of all closed classes ? In the general theory of superposition of functions the following „finiteness question" also plays an essential role: Is it possible to generate by superposition all closed classes of Boolean functions from finite subsets of them ? Answers to these questions were given by E.L.Post ([1],[2]). He characterized all classes of Boolean functions closed under superposition. Post's results were an essential enrichment of the theory of Boolean functions and its applications. The closed classes form a complete lattice with respect to set inclusion, which is countable, atomic and dually atomic. Today it is known that this lattice is algebraic.

Post also solved the completeness problem for Boolean functions. His solution is based on the following ideas. Since the lattice of all closed classes of Boolean functions is dually atomic, a set of Boolean functions is functionally complete if and only if it is not entirely contained in a dual atom. The dual atoms in the lattice of all closed classes of Boolean functions are called maximal or precomplete classes. Post described five maximal classes and he has proved that these five maximal classes are exactly all maximal classes of Boolean functions. They are the following ones:

C_3 (and C_2, resp.): the class of all Boolean functions such that

$$f(0,\dots,0) = 0 \text{ (with } f(1,\dots,1) = 1),$$

L_1: the class of all Boolean functions such that $f(x_1,\dots,x_n) = c_0+$

$c_1x_1+\ldots+c_nx_n$ (+ addition mod 2, $c_i \in \{0,1\}$). These functions are called linear Boolean functions.

A_1: the class of all Boolean functions such that from $x_1 \leq y_1, \ldots, x_n \leq y_n$ it follows $f(x_1,\ldots,x_n) \leq f(y_1,\ldots,y_n)$, these functions are called monotonic Boolean functions,

D_3: the class of all Boolean functions such that $f(Nx_1,\ldots,Nx_n) = Nf(x_1,\ldots,x_n)$, these functions are called self-dual Boolean functions.

This result establishes the great importance of maximal classes. If the maximal classes are considered as sets of term functions of universal algebras with carrier $\{0,1\}$, the main topic of this book, the (two-element) preprimal algebra, is given. Later the concept of a preprimal algebra will be defined more precisely.

Up to now only functions on the two-element set $A = \{0,1\}$ are considered. It is usual to generalize this and to investigate also functions on any finite set. Instead of an arbitrary k-element set where k is an integer, $k > 2$, the set $A = \{0,\ldots,k-1\}$, is often considered. This generalization arose in the development of logic at the beginning of the twentieth century. As in the geometry of the nineteenth century the doubts about the absoluteness of the Euclidian geometry led to the development of non-Euclidian geometries, in logic at the beginning of the twentieth century the doubts about the „tertium non datur" of Aristoteles led to the development of multiple-valued logics. Philosophical problems arising from the idea that there exist statements which are neither true nor false led to the formulation of multiple-valued logics by Łukasiewicz and Post. In 1920 Łukasiewicz ([1]) constructed a three-valued propositional calculus which was generalized in Łukasiewicz/Tarski[1] to a k-valued propositional calculus, $k > 2$, with a single distinguished truth-value.

The k-valued propositional calculus, $k > 2$, with the truth-values $0,\ldots,k-1$ is constructed by means of all functions on the set $A = \{0,\ldots,k-1\}$, the k-valued functions, as propositional functions. In analogy to the Boolean functions in this case closed classes of k-valued functions are considered. The closed classes of k-valued functions form an algebraic, atomic, and dually atomic lattice. But, contrary to the Boolean case little is known about the structure of the lattice of all closed classes of functions on $\{0,\ldots,k-1\}$ if $k>2$. While every closed class of Boolean functions has a finite basis, there are closed classes of functions on $\{0,\ldots,k-1\}$, $k>2$, which cannot be generated by finitely many functions. While there are countably infinitely many closed classes of Boolean functions, the lattice of the closed classes of functions on $A=\{0,\ldots,k-1\}$, $k>2$, has the cardinality of the continuum. As for $k=2$ also for $k>2$ the completeness problem plays a central role in the study of the superposition of functions on $A=\{0,\ldots,k-1\}$. The completeness problem of the k-valued logic can be considered as solved if all maximal classes of functions on $\{0,\ldots,k-1\}$, the dual atoms in the lattice of all closed classes, are known.

At the beginning of the fifties there was a number of mathematicians who have studied this problem, for instance P.S.Novikov, S.V.Jablonskij, and A.V.Kuznezov. For $k=3$ the completeness problem was solved by Jablonskij ([1]) and for an arbitrary $k>3$ by I.G.Rosenberg in [2]. The basic idea in Rosenberg's paper consists in the characterization of all maximal classes of functions on $\{0,\ldots,k-1\}$, with the help of relations. The difficult part of Rosenberg's proof is to show that all maximal classes have been found. The maximal classes of k-valued functions, $k>2$, correspond exactly to six classes O,P,L,E,C,B of relations consisting of order relations, binary relations determined by special permutations, quaternary relations determined by special p-groups, equivalence relations, central relations and h-universal relations, respectively. All maximal classes of functions on A are characterized by these relations, therefore all preprimal algebras $\underline{A}=\langle A,F\rangle$ are determined, where F is a maximal class on A.

At this place it must be mentioned that the research of the closed classes of Boolean and arbitrary k-valued functions received important stimulations from computer science (Hotz [1]). The reason for this is the following: Let basis elements, called gates, be given, each of which has one or several inputs and a single output. The gate receives signals on the inputs and transforms them into the output signal. The input and output signals are elements of $A = \{0,\dots,k-1\}$. The action of a gate can be described by the assignment of the output signal $f(x_1,\dots,x_n)$ to every ordered n-tuple $(x_1,\dots,x_n)$ of input signals. Thus a gate realizes a function $f: A^n \to A$ of n variables ranging in the finite set $A = \{0,\dots,k-1\}$ with values in $A = \{0,\dots,k-1\}$.

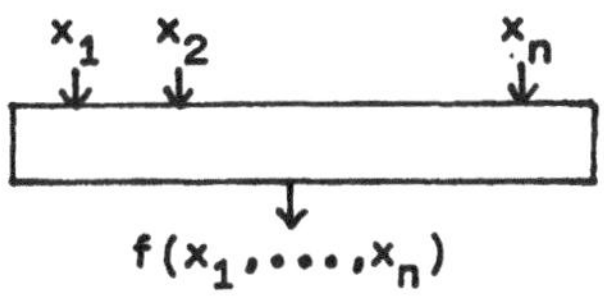

Let a collection of gates realizing functions f_i on A be given. These gates can be combined into switching circuits which realize further functions by attaching outputs of certain gates to inputs of certain gates so that the resulting circuit possesses a single output and no feedback is created. Reordering and identification of inputs is also allowed. This means the single output of the circuit defines a unique function f of the inputs. The function f can be described as a superposition of the f_i's. Consider, e.g., the following gates realizing the Boolean functions $x_1 \wedge x_2$, $x_1 \vee x_2$

The following circuit is a combination of these gates:

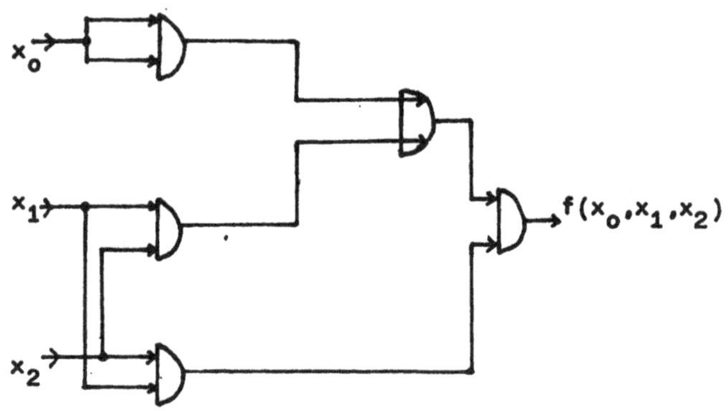

This switching circuit describes the Boolean function with the table

x_0	x_1	x_2	$f(x_0,x_1,x_2)$
0	0	0	0
0	0	1	0
0	1	0	0
0	1	1	1
1	0	0	0
1	0	1	0
1	1	0	0
1	1	1	1

It holds $f(x_0,x_1,x_2) = ((x_0 \vee x_0) \vee (x_1 \wedge x_2)) \wedge (x_1 \vee x_2)$. It is natural to ask whether a given set of gates suffices to construct every switching circuit. This corresponds to the completeness problem of the k-valued logic. Therefore also in this connection it is essential to know the maximal classes of Boolean or arbitrary k-valued functions. From the economical point of view it is essential to construct switching circuits in a optimal way using a minimum of gates. For example, in the above case it is easy to simplify f to $x_1 \wedge x_2$, but f with $f(x_0,x_1,x_2) = ((x_0 \vee x_0) \vee (x_1 \wedge x_2)) \wedge (x_1 \vee x_2)$ and g with $g(x_1,x_2) = x_1 \wedge x_2$ are different functions although they do the same. In order to identify f and g one can consider identities of the finite universal algebra $\underline{A} = \langle A,F \rangle$ with $A = \{0,\dots,k-1\}, k > 1$. This motivates in addition why one investigates algebras $\underline{A} = \langle A,F \rangle$.

The closed class $\mathbf{F}$ which arises from fundamental operations F of an algebra $\underline{A}$ by superposition is called the collection of all term functions of $\underline{A}$ or the clone of operations of $\underline{A}$ (Cohn [1]). It is assumed that F contains all projections, i.e. all functions e_i with $e_i(x_1,\ldots,x_n) = x_i$. Algebras and their identities are also studied in the propositional logic if logically equivalent formulas of the propositional calculus are considered. Let $V_{\underline{A}}$ be the equationally defined class generated by $\underline{A}$, the so-called variety generated by $\underline{A}$. $V_{\underline{A}}$ can be considered as the class of all algebras of the same type as $\underline{A}$ satisfying the same identities as $\underline{A}$. There is a natural correspondence between the closed class of functions on the finite set A generated by F and the variety $V_{\underline{A}}$ generated by $\underline{A} = \langle A,F\rangle$. In fact, $V_{\underline{A}}$ is completely determined by the $V_{\underline{A}}$ -free algebra on a countable number of generators. Since an element of the $V_{\underline{A}}$-free algebra on a countable number of generators may be considered as an n-ary term function on $\underline{A}$ for some n, there exists a one-to-one map from the variety generated by $\underline{A} = \langle A,F\rangle$ to the closed class generated by F. These ideas are a basis to apply concepts, theorems, and methods of Universal algebra in the investigation of algebraic properties of k-valued logics or in the theory of switching circuits.

It is natural to consider at first such finite algebras $\underline{A} = \langle A,F\rangle$ and the varieties $V_{\underline{A}}$ in which all functions defined on A are term functions. For instance, the algebras $\langle\{0,1\},\wedge,N\rangle$ of the type (2,1) and $\langle\{0,1\},\wedge,\vee,N\rangle$ of the type (2,2,1) are algebras whose term functions are exactly all Boolean functions. Algebras with distinguished sets of fundamental operations, but with the same collection of term functions, such as $\langle\{0,1\},\wedge,N\rangle$ and $\langle\{0,1\},\wedge,\vee,N\rangle$, are called equivalent. (We often will make no difference between equivalent algebras.)

The algebra $2_{\underline{B}} = \langle\{0,1\},\wedge,\vee,N\rangle$ and the variety generated by $2_{\underline{B}}$ describe the classical propositional calculus since from $\langle\wedge,\vee,N\rangle$ all functions corresponding to the functions of the classical propositional calculus can be obtained by superposition. $2_{\underline{B}}$ generates the variety of Boolean algebras. A finite non-trivial algebra $\underline{A} = \langle A, \wedge, \vee, N \rangle$ is called to be a Boolean algebra if the following identities (B 1) - (B 5) are satisfied:

(B 1) $x \wedge y = y \wedge x,$ $\qquad x \vee y = y \vee x,$

(B 2) $x \wedge (y \wedge z) = (x \wedge y) \wedge z,$ $\qquad x \vee (y \vee z) = (x \vee y) \vee z,$

(B 3) $x \wedge (x \vee y) = x,$ $\qquad x \vee (x \wedge y) = x,$

(B 4) $x \wedge (y \vee z) = (x \wedge y) \vee (x \wedge z),$ $\qquad x \vee (y \wedge z) = (x \vee y) \wedge (x \vee z),$

(B 5) $x \wedge (y \vee Ny) = x,$ $\qquad x \vee (y \wedge Ny) = x.$

(It is not hard to verify that $2_{\underline{B}}$ is a Boolean algebra.)

The property that every Boolean function is a term function of $2_{\underline{B}}$ characterizes $2_{\underline{B}}$ up to equivalence because it holds: A two-element algebra is $2_{\underline{B}}$ (up to equivalence) if and only if every function on $\{0,1\}$ is a term function of $2_{\underline{B}}$. This characterization of the two-element Boolean algebra is of great interest both for the propositional logic and for the theory of switching circuits. But the algebraist endeavours a characterization in algebraic terms, for instance using the properties of the congruence lattice, of the subalgebra lattice, or of the automorphism group. The two-element algebra $2_{\underline{B}} = \langle \{0,1\}, \wedge, \vee, N \rangle$ has no proper non-trivial subalgebras, moreover $\underline{A}$ is simple and has no non-identical automorphisms . It is not hard to prove that the congruence lattices of all algebras of $V_{\underline{A}}$ are distributive and the congruences are pairwise permutable. In 1964 A.F.Foster and A.L.Pixley showed that these properties suffice for a two-element algebra to be equivalent to the two-element Boolean algebra. Thus we have a universal algebraic completeness criterion for Boolean functions. Does this criterion also hold for any finite algebra whose term functions are exactly all functions defined on the carrier of the algebra ? Such algebras have been investigated by Foster since 1953 and were later denoted as primal algebras. In Foster/Pixley [1], [2] it is proved that for any primal algebra the same algebraic characterization is satisfied as for the two-element Boolean algebra, i.e. a finite non-trivial algebra $\underline{A} = \langle A, F \rangle$ is primal if and only if

(a) $\underline{A}$ has no proper subalgebra,

(b) $\underline{A}$ has no non-identical automorphisms,

(c) $\underline{A}$ is simple, and

(d) $\underline{A}$ generates a congruence distributive and congruence permutable variety (this means that any algebra in $V_{\underline{A}}$ has a distributive congruence lattice and pairwise permutable congruences).

For instance, using this criterion it can be proved that the k-element algebra $\underline{A} = \langle \{0,\dots,k-1\}, \vee, \wedge, C, D_1,\dots,D_{k-1}, 0,\dots,k-1 \rangle$, $k>2$, whose fundamental operations are defined by $i \vee j = \max(i,j)$, $i \wedge j = \min(i,j)$, $i,j \in \{0,\dots,k-1\}$, $D_i(j) = \begin{cases} k-1 & \text{if } i \leq j \\ 0 & \text{if } i > j \end{cases}$, $C(i) = \begin{cases} k-1 & \text{if } i=0 \\ 0 & \text{if } i>0 \end{cases}$, is primal. This algebra, the so-called Post algebra of the order k, plays the same role for the k-valued propositional calculus, $k>2$, as the two-element Boolean algebra for the two-element propositional calculus since the term functions of the Post algebra of the order k are the functions of the k-valued propositional calculus. Traczyk ([1]) investigated the variety generated by the k-element Post algebra of order k and gave a description by axioms.

By a result of Birkhoff [1] the variety generated by any finite non-trivial algebra $\underline{A}$ coincides with the class of all homomorphic images of subalgebras of direct powers of $\underline{A}$ ($V_{\underline{A}}$ = HSP $\underline{A}$). For a primal algebra, however, there is a much simpler representation of any algebra $\underline{B} \in V_{\underline{A}}$. An arbitrary Boolean algebra is isomorphic to a subdirect power of the two-element Boolean algebra. It follows that all identities which are satisfied in any Boolean algebra, are satisfied in the two-element Boolean algebra, i.e. the two-element Boolean algebra is equationally maximal. These and other important properties of Boolean algebras transfer to varieties generated by a primal algebra since the variety generated by a primal algebra is isomorphic as a category to the category of Boolean algebras,(with homomorphisms as morphisms), conversely, any variety isomorphic to Boolean algebras is generated by a primal algebra (T.-K.Hu [1]). Therefore primal algebras and the varieties generated by primal algebras have many „good" properties. Which of these properties are preserved if the primality is weakened ? A.L.Foster, A.F.Pixley, R.W.Quackenbush and other algebraists considered finite algebras as generalizations of primal algebras such that not every function on A is a term function of $\underline{A}$ but every function which satisfies a certain property is a term function. The requirement that all functions which preserve all subalgebras of $\underline{A}$ are term functions of $\underline{A}$, leads to semiprimal algebras. If every function which preserves all congruences on $\underline{A}$ is a term function of $\underline{A}$, one obtains hemiprimal algebras. Other generalizations of primal algebras are demiprimal, demi-semi-primal, and quasiprimal algebras. (The meaning of „preservation" will be explained later.)

The investigation of preprimal algebras, (i.e. of such finite algebras whose term functions form a maximal class), is motivated by the great importance of maximal classes of functions on a finite set for the solution of the completeness problem of k-valued logic, but also by the fact that the set of term functions of preprimal algebras constitutes the stage level below the term functions of primal algebras in the lattice of all closed classes. Single preprimal algebras were studied by Knoebel ([2]and [3]) and by Froemke ([2]). In the paper of Froemke ([2]) the notion „preprimal algebra" is used for the first time. The first attempts of a systematic treatment of preprimal algebras and of the varieties generated by them can be found in Knoebel ([5]) and in the author's thesis (Denecke[1]) who was inspired by H.J.Hoehnke. Further contributions are Denecke ([2],[4],[5]). (More about the short history of maximal classes and of preprimal algebras can be found in 4.1).

The above mentioned theorem by Foster/Pixley gives an algebraic characterization of primal algebras. A main problem treated in this book is the search for an algebraic characterization of preprimal algebras. This problem will be solved for some classes of preprimal algebras. Let us demonstrate by an example what we have in mind with an algebraic characterization of a preprimal algebra (Foster/Pixley[1]). A finite non-trivial algebra $\underline{G} = \langle G,+,\cdot\rangle$ of the type (2,2) is said to be a double group of the order k, $k > 2$, if the following conditions are satisfied:

(DG 1) $\langle G,+\rangle$ is the cyclic group of the order k. (Write 0 for the identity element of +, let e be a fixed generator of $\langle G,+\rangle$ and let G' denote G with 0 deleted).

(DG 2) $\langle G',\cdot\rangle$ is a fixed group of order k-1 and has e as its identity.

(DG 3) $0 \cdot x = x \cdot 0 = 0$ for all $x \in G$.

All functions on G which preserve the set $\{0\}$ are term functions of the algebra $\underline{G}$, $\{0\}$ is the only subalgebra. $\underline{G}$ is a preprimal algebra since $\langle +, \cdot\rangle$ generates a maximal class and $\underline{G}$ has the following properties:

(a') $\underline{G}$ has exactly one proper subalgebra which is a one-element algebra,

(b) $\underline{G}$ has no proper automorphisms,

(c) $\underline{G}$ is simple,

(d) $\underline{G}$ generates a congruence distributive and congruence permutable variety.

But, moreover, the converse is also valid, i.e. $\underline{G}$ is (up to equivalence) characterized by these properties. Not only $\underline{G}$ but also a series of other preprimal algebras, (which are not equivalent to $\underline{G}$, e.g., all algebras which generate arithmetical varieties), can be characterized by similar algebraic conditions.

In the following we briefly describe the content of the chapters of this book. In chapter 1 the basic concepts from the theory of superposition of functions on a finite set and from Universal algebra are explained. Here also quasi-, semi-, hemi-, para-, and cryptoprimal algebras, i.e. the generalizations of primal algebras which were mentioned above are defined. Chapter 2 is concerned with arithmetical varieties of algebras. From Pixley's characterization of quasiprimal algebras ([3])

characterizations of semiprimal, demiprimal, and primal algebras are derived. In 2.3 those properties of hemiprimal algebras are given which are used in the following chapters. Chapter 3 develops an algebraic characterization of the two-element preprimal algebras. From this characterization conditions can be derived for the primality and for weaker completeness properties of two-element algebras. Using Post's classification of all closed classes of Boolean functions results are given about all two-element algebras and the varieties generated by two-element algebras. In this connection a new, shorter proof of Lyndon's ([1]) theorem, that any two-element algebra is finitely axiomatizable, is presented. Chapter 4 describes all forms of preprimal algebras $\underline{A} = \langle A, F \rangle$, card $A > 2$, by using Rosenberg's classification of all maximal classes of k-valued functions as starting point. One of the main results is the following theorem on preprimal algebras: A preprimal algebra is either quasiprimal or constantive. Further results given in chapter 4 are: Every proper non-trivial homomorphic image and every proper non-trivial subalgebra of a preprimal algebra is primal. The automorphism group of a preprimal algebra is cyclic and of prime order. The investigation of independent sets of preprimal algebras in the sense of E.Marczewski leads to the result that an independent set in a preprimal algebra has at most one element. In this connection all free preprimal algebras are determined as well.

Preprimal algebras are investigated up to equivalence only. The type of the algebras under consideration plays no significant role since the classification by Rosenberg involves term functions rather than fundamental operations. Therefore preprimal algebras are investigated up to equivalence only and the reader's attention will be directed to the study of weak automorphisms of preprimal algebras in the sense of Goetz and Marczewski. The main result is that for an arbitrary preprimal algebra the group of all weak automorphisms is isomorphic to the group of all clone automorphisms. For special classes of preprimal algebras the group of all weak automorphisms is described more precisely. In this connection also all these preprimal algebras are listed in which every permutation on A is an invertible (one-to-one) unary term function of $\underline{A} = \langle A,F \rangle$. Another algebraic aspect of maximal classes was mainly developed by D.Lau in [1], [2], [3]. Lau investigated maximal function algebras (in the sense of A.I.Malcev [5]) and their order (Lau [2]). In 4.4 we present these results and prove, that preprimal algebras have finite types (with one exception).

In chapter 5 varieties generated by preprimal algebras are discussed. The first step in the direction of an algebraic characterization of preprimal algebras is a characterization of those preprimal algebras which generate arithmetical varieties. The investigations in 5.3 and 5.4 result in a complete survey of all preprimal algebras generating congruence permutable varieties and an almost complete survey of all preprimal algebras generating congruence distributive varieties. Only for one class of preprimal algebras the congruence distributivity is an open problem. In 5.4 it is proved that an algebra generating a 2-distributive variety, (i.e., if a majority function d with $d(x,x,y) = d(x,y,x) = d(y,x,x) = x$ is a term function), is preprimal if and only if the direct power $\underline{A}^2$ has (up to isomorphisms) exactly one subalgebra which is distinct from the diagonal subalgebra. If this only subalgebra of $\underline{A}^2$ is rectangular (Pixley [7]) one obtains a further characterization of preprimal algebras generating arithmetical varieties. In 5.5 the subdirectly irreducible algebras in a variety generated by a preprimal algebra are discussed. It is known that the generating algebra is the only subdirectly irreducible algebra if it is primal. For varieties generated by a preprimal algebra (with exception for one class) we shall see (using a result of Knoebel [4]) that they contain only finitely many subdirectly irreducible algebras, all of them are finite. The knowledge of the subdirectly irreducible algebras allows to state structure theorems for the varieties corresponding to five of the six classes of preprimal algebras. Moreover, using the subdirectly irreducible algebras the lattice of the subvarieties of the variety generated by a preprimal algebra can be described. A result of Foster ([3]) is applied to determine the lattice of all subvarieties in case of one class of preprimal algebras. Results of Knoebel ([4]) are presented for another class. In the last section of chapter 5 it is proved that the preprimal algebras from 4 of the 6 classes are finitely axiomatizable. This problem is open in the remaining cases.

In chapter 6 two different completeness properties of universal algebras are discussed, primality and functionally completeness of algebras. We use the developed properties of preprimal algebras in order to give new and shorter proofs of well-known primality and functionally completeness conditions for universal algebras. In this connection the presumptions are given under which a primal algebra $\underline{A}^+$ arises from a functionally complete algebra $\underline{A}$ by addition of exactly one nullary function which is not a term function of $\underline{A}$. In 6.3 primal and preprimal algebras with a single fundamental operation are considered. G. Rousseau ([1]) obtained the unexpected result that for primal algebras with a single fundamental operation (which must be at least binary) condition (d) in the algebraic characterization of primal algebras is redundant. Rousseaus's original proof of this theorem can be shortened by a method of Knoebel ([4]). Using Rousseau's theorem an algebraic characterization of preprimal algebras with a single fundamental operation can be derived (6.2 - Theorem 5).

For the investigations treated in this book the classification of all maximal classes by Rosenberg plays a fundamental role. Therefore in 6.4 a sketch of Quackenbush's new proof of Rosenberg's theorem is given using algebraic results. The complete proof is contained in the appendix.

Examples for preprimal algebras are certain BCK-algebras. They originate from the classical and non-classical propositional calculi and are described in chapter 7. The chapters 2,3,4, and 5 are supplemented by problems and complements which arise from the general theory and which can stimulate the reader for further research.

1. Basic concepts

In this chapter we will briefly review some basic concepts of the theory of superposition of functions on finite sets and of Universal algebra.

1.1 Functions on a finite set

Let $A = \{0,\dots,k-1\}$ and let k be a natural number, $k > 2$. An n-ary function on A is a function $f: A^n \longrightarrow A$. Let $F_n(A)$ be the set of all n-ary functions: $F_n(A) = \{f: A^n \longrightarrow A\}$ and let $\underline{F}(A)$ be the class of all functions defined on A, i.e. $\underline{F}(A) = \bigcup_{n=0}^{\infty} F_n(A)$. We give some examples for functions on A: $x_1 + x_2 (\mathrm{mod}\, k)$, $\max(x_1,x_2)$, $\min(x_1,x_2)$, $(k,x_1,x_2$ integers, max and min with respect to the ordinary order relation). The functions $e_i^n \in F_n(A)$ with $e_i(x_1,\dots,x_n) = x_i$ for all $i = 1,\dots,n$, the so-called projections, are examples which will appear frequently in the sequel. Let F be a subset of $\underline{F}(A)$, $F \subseteq \underline{F}(A)$. By the result of a superposition over F we will understand a function from $\underline{F}(A)$ which is obtained from the functions of F by applying the following operations in any finite number of times:

- permutation of variables
- identification of variables
- substitution of a variable in the arguments of a function by a function.

The class of all superpositions of functions of F is called $\bar{F}$. All classes $F \subseteq \underline{F}(A)$ with $\bar{F} = F$ are called closed classes with respect to superposition. If for a subset F_1 of a closed class F it holds $\bar{F}_1 = F$, i.e. the set of all superpositions of F_1 is the class F, then F_1 is called a generating system of F. A generating system F_1 of F is said to be a basis of F if no proper subset of F_1 is a generating system of F. A class F with $\bar{F} = \underline{F}(A)$ is called functionally complete in $\underline{F}(A)$. For instance, the system $\langle 0,\dots,k-1, \max(x_1,x_2), \min(x_1,x_2), j_i (0 \leq i \leq k-1) \rangle$ with $j_i(x) = \begin{cases} k-1 & \text{if } x=i \\ 0 & \text{if } x \neq i \end{cases}$ is functionally complete. In this system

0,...,k-1 are nullary functions; max, min are binary functions, and the j_i are unary functions (Jablonskij [1]). Another well-known example for a functionally complete set in $\underline{F}(A)$ is the system $\langle \max(x_1,x_2), x_1+x_2 (\bmod k)\rangle$. A function $f \in F_n(A)$ is said to be Sheffer in $\underline{F}(A)$ if $\{f\}$ is functionally complete in $\underline{F}(A)$. Thus f is Sheffer if and only if each function of $\underline{F}(A)$ can be expressed by superposition of f, only. The function $\max(x_1,x_2)+1(\bmod k)$ is an example for a Sheffer function, (k,x_1,x_2 natural numbers).

A closed class $M \subseteq \underline{F}(A)$ is said to be maximal (or precomplete) in $\underline{F}(A)$ if $\bar{M} \neq \underline{F}(A)$, however for any function $f \in \underline{F}(A)$, $f \notin \bar{M}$, it holds $\overline{M \cup \{f\}} = \underline{F}(A)$. If M is a maximal class in $\underline{F}(A)$, M cannot be properly extended to a closed proper subclass of $\underline{F}(A)$. The set of all closed classes from $\underline{F}(A)$ forms an algebraic lattice with respect to the set-theoretical inclusion. Historically this lattice was first investigated by E.L.Post ([1],[3]) for k=2. Post proved that each closed class in $\underline{F}(\{0,1\})$ has a finite basis. For k=2 the lattice of all closed classes is countable infinite. For $k > 2$ there exist closed classes possessing no finite basis. The lattice of all closed classes has the cardinality of the continuum. This difference between k=2 and $k > 2$ signalizes the fact that the propositional calculi of multiple-valued logics ($k > 2$) are intrinsically more complex than the two-valued one.

The maximal classes are the dual atoms of the lattice of all closed classes. Using the existence of a finite complete system and Zorn's Lemma it can be shown that the lattice of all closed classes is dually atomic, i.e. that each proper closed class can be extended to a maximal one. The set of all maximal classes is finite (Jablonskij [1]). Therefore a set of functions $F \subseteq \underline{F}(A)$ is functionally complete if and only if F is not a subset of a maximal class, i.e. F is functionally complete if and only if for every maximal class in $\underline{F}(A)$ the set F contains a function f not belonging to the maximal class. This is the most general completeness criterion. In chapter 6 we will derive other completeness propositions from this completeness criterion. The functional completeness problem is solved by determining of all maximal classes. Post determined all maximal classes in $\underline{F}(\{0,1\})$, Jablonskij in $\underline{F}(A)$ for card A = 3 (Jablonskij [1]) and Rosenberg gave a complete classification of all maximal classes in $\underline{F}(A)$ for $k > 3$ by relations (Rosenberg [2]).

Let $k \geq 2$ and $1 \leq h \leq k$. An h-ary relation on $A = \{0,\dots,k-1\}$ is a subset ϱ of the cartesian power $A^h = A \times \dots \times A$. The cartesian power ϱ^n of the relation ϱ is a relation defined on A^n in the following manner: Let $\tilde{a}_i = (a_{1i},\dots,a_{ni}) \in A^n$, then $(\tilde{a}_0,\dots,\tilde{a}_{h-1}) \in \varrho^n$ if and only if $(a_{j0},\dots,a_{jh-1}) \in \varrho$ for all $j=1,\dots,n$. We say that a function $f \in F_n(A)$ preserves ϱ if and only if $(f(\tilde{a}_0),\dots,f(\tilde{a}_{h-1})) \in \varrho$ whenever $(\tilde{a}_0,\dots,\tilde{a}_{h-1})$ belongs to ϱ^n. In other words, f preserves ϱ if and only if for every matrix of type n x h whose rows are all contained in ϱ, the column values of f form an h-tuple belonging also to ϱ. In the terminology of Universal algebra: f preserves ϱ if and only if $\langle \varrho, f \rangle$ is a subalgebra of $\langle A^h, f \rangle$.

Let Pol ϱ be the set of all $f \in \underline{F}(A)$ preserving ϱ. For a set R of relations on A we define PolR by Pol R $= \{ f \in \underline{F}(A) \mid f$ preserves every $\varrho \in R\}$. The set Pol R is a closed class containing all projections e_i^n. On the other-hand side, if F is a closed class containing all projections there exists an h-ary relation R with F = Pol R. For an arbitrary class $F \subseteq \underline{F}(A)$ we define: Inv F $= \{ \varrho \mid \varrho$ is a relation on A which is invariant with respect to every function $f \in F\}$. The operators Pol and Inv define a Galois-connection between the subsets of $\underline{F}(A)$ and the subsets of $\underline{R}(A)$, where $\underline{R}(A)$ is the set of all h-ary relations on A for all h with $1 \leq h \leq k$. For an extensive treatment of relations and the functions they preserve, see Pöschel/Kalužnin ([1]).

Let ι_h be the set of all ordered h-tuples defined by
$\iota_h = \{(a_0,\dots,a_{h-1}) \in A^h \mid i,j \in \{0,\dots,h-1\}$ and $a_i = a_j$ for $i \neq j\}$

For $\varrho \supseteq \iota_h$ an h-ary relation ϱ will be called totally reflexive. For h=2 this means that ϱ is reflexive, i.e. $(a,a) \in \varrho$ for all $a \in A$, in other words, the diagonal $\Delta(A)$ is included in ϱ.

We denote a unary (constant) function from $\underline{F}(A)$ by c_i ($c_i(x) \equiv i$ for all $x \in A$ and for all i, $(0 \leq i \leq k-1)$). The constant function c_i preserves ϱ if and only if $(i,\ldots,i) \in \varrho$. It follows that, if ϱ is an h-ary totally reflexive relation and $2 \leq h \leq k$, then all constant functions c_i, $(0 \leq i \leq k-1)$, belong to Pol ϱ

The n-ary function f essentially depends on the i-th variable $(1 \leq i \leq n)$ if there are n-tuples $\tilde{a} = (a_1,\ldots,a_{i-1},b,a_{i+1},\ldots,a_n)$, $\tilde{a}' = (a_1,\ldots,a_{i-1},c,a_{i+1},\ldots,a_n)$ in A^n such that $f(\tilde{a}) \neq f(\tilde{a}')$. In this case the ith variable is said to be essential for f.

1.2 Finite algebras

A finite algebra is a pair $\underline{A} = \langle A,F \rangle$ where A is a finite set and F is a set of functions on A. The elements of F are called fundamental operations. We do not distinguish between isomorphic algebras and assume that the k-element set is identical with $\{0,\dots,k-1\}$. F is a subset of $\underline{F}(A)$. If k=1, we call $\underline{A}$ a trivial algebra. If the set $\bar{F}$ of all superpositions of F contains all projections, then $\bar{F}$ by definition is the set of all term functions of $\underline{A}$. The set $\bar{F}$ of all superpositions of F is also called clone of operations of $\underline{A}$ (Cohn [1]). For many investigations not only the fundamental operations are essential, but we are also interested in all term functions of the algebra $\underline{A}$. Therefore we will often make no difference between two algebras with the same carrier A and the same collection of term functions. We define: Two algebras $\underline{A}_1$ and $\underline{A}_2$ are called equivalent if they possess the same elements and the same collection of term functions, i.e. $\underline{A}_1 = \langle A,F_1 \rangle$ and $\underline{A}_2 = \langle A,F_2 \rangle$ are equivalent if and only if $\bar{F}_1 = \bar{F}_2$. (The algebraic theories generated by equivalent algebras are equal.)

In the language of Universal algebra the concept of functional completeness corresponds with primality. A finite non-trivial algebra $\underline{A}$, $\underline{A} = \langle A,F \rangle$, where F is functionally complete in $\underline{F}(A)$, is called primal. Every function from $\underline{F}(A)$ is a term function of the primal algebra $\underline{A} = \langle A,F \rangle$. Primal algebras are equivalent to the basic-k-element Post algebra $\underline{k}$ for $k \geq 3$ and they are equivalent to the 2-element Boolean algebra for k=2.

An element from $\overline{F \cup \{c_i \mid 0 \leq i \leq k-1\}}$, ($c_i$ constant functions on A), is said to be a polynomial function of the algebra $\underline{A}$. A concept closely related to the primality is the functional completeness of a k-element algebra. A finite non-trivial algebra $\underline{A}$ is functionally complete if and only if every function on A is a polynomial function of the algebra $\underline{A}$, i.e. $\overline{F \cup \{c_i \mid 0 \leq i \leq k-1\}} = \underline{F}(A)$. In Rosenberg ([5]) the set $F \subseteq \underline{F}(A)$

with $\overline{F \cup \{c_i \mid 0 \leqslant i \leqslant k-1\}} = \underline{F}(A)$ is called complete with constants. In Universal algebra the completeness with constants of a set of functions corresponds to the functional completeness of an algebra. Thus it is a difference between the functional completeness of a set of functions and the functional completeness of an algebra.

A finite non-trivial algebra $\underline{A} = \langle A,F\rangle$ is called constantive if all constant functions on A are term functions of $\underline{A}$, i.e. $\{c_i \mid 0 \leqslant i \leqslant k-1\} \subseteq F$. Let $\underline{A}^+$ be the algebra obtained from $\underline{A}$ by making each element of $\underline{A}$ the value of a new constant term function. $\underline{A}^+$ has, as a constantive algebra, no proper subalgebras and no non-identical automorphisms. The congruence relations of $\underline{A}^+$ and $\underline{A}$ agree together. Clearly, an algebra $\underline{A}$ is functionally complete if and only if $\underline{A}^+$ is primal.

If we consider a class of algebras of the same type (as usual) then we can define homomorphisms, isomorphisms, automorphisms, subalgebras, direct and subdirect products, free algebras, and so on. Further it is well-known that the set of all congruence relations of an algebra $\underline{A}$ forms a complete algebraic lattice. An algebra is called simple if this lattice consists of the zero-congruence and the universal congruence, only. For the basic concepts of Universal algebra we refer to Grätzer's book (Grätzer[1]).

In accordance with Pixley ([8]) instead of Grätzer's „polynomial symbol", „polynomial", and „algebraic function" we use „term", „term function", and „polynomial function", respectively. A term of the type τ is simply a term (in the ordinary logical sense) in first order language of type τ .

Now we direct our attention to generalizations of primal algebras. We obtain weaker concepts of the algebraic completeness requiring that all functions which preserve certain well-defined algebraic properties, are term functions of the algebra. In this way we obtain quasiprimal, semiprimal, demiprimal, demi-semi-primal and hemiprimal algebras (Foster[8], Foster/Pixley[1], [2], Pixley[4], [5], [7], [8], Werner[2]). A good review is given in Werner ([2]).

<u>Definition 1:</u> A finite non-trivial algebra $\underline{A} = \langle A,F \rangle$ is semiprimal if every function from $\underline{F}(A)$ preserving all subalgebras of $\underline{A}$ is a term function of $\underline{A}$.

<u>Definition 2:</u> A finite non-trivial algebra $\underline{A} = \langle A,F \rangle$ is demiprimal if $\underline{A}$ has no proper subalgebras and every function from $\underline{F}(A)$ preserving all automorphisms of $\underline{A}$ is a term function of $\underline{A}$.

<u>Definition 3:</u> A finite non-trivial algebra $\underline{A} = \langle A,F \rangle$ is quasiprimal if every function from $\underline{F}(A)$ preserving all subalgebras of $\underline{A}$ and all isomorphisms between non-trivial subalgebras of $\underline{A}$ is a term function of $\underline{A}$.

<u>Definition 4:</u> A finite non-trivial algebra $\underline{A} = \langle A,F \rangle$ is hemiprimal if every function $f \in \underline{F}(A)$ preserving all congruence relations on $\underline{A}$ is a term function of $\underline{A}$.

The meaning of „preservation" is the following one: A function $f \in F_n(\underline{A})$ preserves

- the subalgebra $B \subseteq A$ if for any $a_1,\dots,a_n \in B$ we have $f(a_1,\dots,a_n) \in B$
- the isomorphism φ between non-trivial subalgebras of $\underline{A}$ if $\varphi(f(a_1,\dots,a_n)) = f(\varphi(a_1),\dots,\varphi(a_n))$
- the congruence relation Θ on $\underline{A}$ if for any $a_i,b_i \in A,(i=1,\dots,n)$, from $(a_1,b_1) \in \Theta,\dots,(a_n,b_n) \in \Theta$ it follows that $(f(a_1,\dots,a_n),f(b_1,\dots,b_n)) \in \Theta$.

Therefore the set of all term functions of quasi-, semi-, demi-, and hemiprimal algebras can be written as Pol Q $= \{ f \in \underline{F}(A)\ f \mid$ preserves all

$\rho \in Q\}$. Here Q is a set of relations on A. For semiprimal algebras Q is a set of unary relations, and for quasi-, demi-, and hemiprimal algebras a set of binary relations. We say that an algebra $\underline{A}$ preserves a relation ρ if every term function of $\underline{A}$ preserves ρ and every function preserving ρ is a term function of $\underline{A}$; in this case we have $F = \{f \mid f$ preserves $\rho\}$. We write $\underline{A}_\rho$ for the algebra just described. From the definitions we obtain the following proposition: Every demiprimal algebra is quasiprimal and every semiprimal algebra is also quasiprimal.

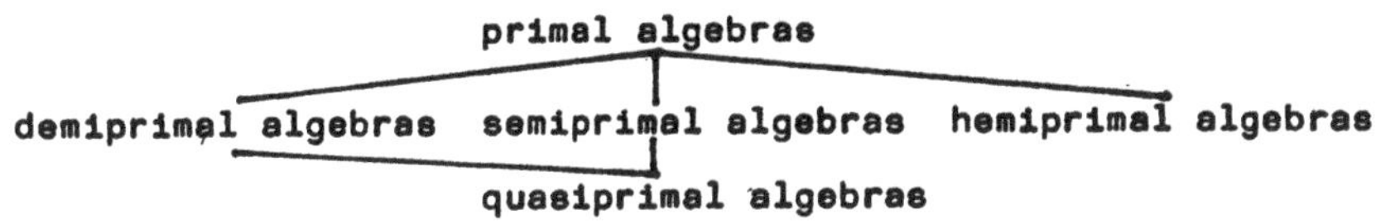

Using the properties of the congruence lattices algebras can be classified in another way. The congruence lattice of $\underline{A}$ is said to be distributive if for any congruences $\theta_0, \theta_1, \theta_2$ on $\underline{A}$ the following identity is satisfied: $\theta_0 \vee (\theta_1 \wedge \theta_2) = (\theta_0 \vee \theta_1) \wedge (\theta_0 \vee \theta_2)$. The congruences on $\underline{A}$ are pairwise permutable if for any two congruences θ_0, θ_1 it holds $\theta_0 \cdot \theta_1 = \theta_1 \cdot \theta_0$, where $\cdot$ denotes the relational product.

Definition 5: A finite non-trivial algebra $\underline{A} = \langle A,F \rangle$ is arithmetical if the congruence lattice of $\underline{A}$ is distributive and all congruences are pairwise permutable.

While quasiprimal, semiprimal, demiprimal, and arithmetical algebras are defined in the terminology of Universal algebra, the concept of a preprimal algebra originates from the theory of superposition of functions. In the following chapters we will investigate the connections between preprimal algebras and quasiprimal, semiprimal, demiprimal, and hemiprimal algebras.

Definition 6: A finite non-trivial algebra $\underline{A} = \langle A,F \rangle$ is preprimal if F is a maximal class in $\underline{F}(A)$.

1.3 Varieties

Varieties are classes of algebras which can be characterized by collections of identities. The theory of varieties is equivalent to the theory of systems of identities. Let $\underline{p}$ and $\underline{q}$ be terms of a given type τ, then $\underline{p} = \underline{q}$ is called an identity in a class of algebras K if the induced term functions p and q are identical in each algebra of the class K, or equivalently $p(x_1,\dots,x_n) = q(x_1,\dots,x_m)$ if $x_1,\dots,x_n,\dots,x_m \in A$, $\underline{A} \in K, (n < m)$. Let Id $\underline{A}$ be the system of all identities in $\underline{A}$. An algebra $\underline{A}$ is said to be equationally maximal if $\underline{A}$ is non-trivial and for all algebras $\underline{B}$ of the same type as $\underline{A}$ from Id $\underline{B} \supseteq$ Id $\underline{A}$ it follows Id $\underline{B}$ = Id $\underline{A}$. The class of all algebras $\underline{B}$ of the same type as $\underline{A}$ such that Id $\underline{B} \supseteq$ Id $\underline{A}$ is called the variety generated by $\underline{A}$ and will be denoted by $V_{\underline{A}}$. In other words, $B \in V_{\underline{A}}$ if and only if $\underline{B}$ has the same type as $\underline{A}$ and all identities of $\underline{A}$ are also identities of $\underline{B}$.

If we denote by P $\underline{A}$ the class of all direct powers of $\underline{A}$, by S $\underline{A}$ the class of all subalgebras of $\underline{A}$, by H $\underline{A}$ the class of all homomorphic images of $\underline{A}$, then it holds, that $V_{\underline{A}}$ = HSP $\underline{A}$, equivalently, each algebra in the variety generated by $\underline{A}$ can be represented as a homomorphic image of a subalgebra of a direct power of $\underline{A}$ (Birkhoff[1]). If $\underline{A}$ is equationally maximal, the variety generated by $\underline{A}$ is said to be minimal.

We assume that the reader is familiar with the notions of a subdirect product of algebras, of subdirectly irreducible algebras and with the Birkhoff's theorem that every algebra is isomorphic to a subdirect product of subdirectly irreducible algebras.

In our further considerations the case plays an essential role where the congruence lattices of all algebras in the variety generated by a finite algebra are distributive (or where the congruences are pairwise permutable). In this case the variety is called congruence distributive (or congruence permutable). Now we pay attention to two definitions of algebras which generate congruence distributive and congruence permutable varieties respectively.

Definition 7: A non-trivial finite algebra $\underline{A} = \langle A,F \rangle$ is cryptoprimal if $\underline{A}$ is simple, has no proper subalgebra, and generates a congruence distributive variety (Bacsich [1]).

Definition 8: A non-trivial finite algebra is paraprimal if each subalgebra of $\underline{A}$ is simple and $\underline{A}$ generates a congruence permutable variety (Clark/Krauss[1]).

Congruence distributive and congruence permutable varieties can be characterized by Malcev-type-conditions.

Theorem 1 (A.I.Malcev): The congruences on every algebra of a variety V are pairwise permutable if and only if there exists a ternary term $\underline{p}$ in V such that for every algebra $\underline{A}$ in V and for all $x,y \in A$:
$p(x,x,y) = y, \quad p(x,y,y) = x.$

In [1] B.Jonsson gave a necessary and sufficient Malcev-type-condition for a variety being congruence distributive.

Theorem 2: A variety V is congruence distributive if and only if there is a natural number $n \geq 2$ and ternary terms $\underline{t}_0,\ldots,\underline{t}_n$ in V such that for every algebra $\underline{A}$ in V, for all $i=0,\ldots,n-1$, and for all $x,y,z \in A$ the following identities are satisfied:

$\underline{t}_n(x,y,z) = x,$

$\underline{t}_i(x,x,z) = \underline{t}_{i+1}(x,x,z) \qquad (i \text{ even}),$

$\underline{t}_i(x,y,y) = \underline{t}_{i+1}(x,y,y) \qquad (i \text{ odd}),$

$\underline{t}_n(x,y,z) = z,$

$\underline{t}_i(x,y,x) = x.$

(Sometimes then we will say that V is n-distributive.)

For n=2 we obtain the identities:

$\underline{t}_1(x,x,y) = x$, $\underline{t}_1(x,y,x) = x$, $\underline{t}_1(x,y,y) = y$.

This part of Theorem 2 was proved by Pixley in [1]:

<u>Corollary 3:</u> If in a variety V there exists a ternary term $\underline{d}$ (the majority term) such that for every algebra $\underline{A}$ in V and for all $x,y \in A$

$\underline{d}(x,x,y) = x$, $\underline{d}(x,y,y) = y$, $\underline{d}(x,y,x) = x$,

then V is congruence distributive (more exactly 2-distributive).

If an algebra $\underline{A}$ does generate a 2-distributive variety, we say that $\underline{A}$ is 2-distributive. If V is congruence distributive and congruence permutable, then V is said to be arithmetical. From Theorem 1 and Corollary 3 it follows:

<u>Corollary 4:</u> A variety V is arithmetical if and only if there exists a term $\underline{g}$ in V such that for every algebra $\underline{A}$ in V and for all x,z in A

$\underline{g}(x,x,z) = \underline{g}(z,x,x) = \underline{g}(z,x,z) = z$.

Let $\underline{A} = \langle A,F\rangle$ be an algebra. The element $a \in A$ is called idempotent if $\langle \{a\},F\rangle$ is a subalgebra of $\underline{A}$. The following theorem of B.Csákány ([1]) characterizes varieties whose algebras have no idempotent elements.

<u>Theorem 5:</u> For any variety V the following conditions are equivalent:

(i) None of the algebras in V, consisting of at least two elements, have idempotent elements.

(ii) No algebra in V has a proper subalgebra whose carrier is a class of some congruences on that algebra.

(iii) There exist a natural number n and ternary terms $\underline{p}_1,\dots,\underline{p}_n$ as well as unary terms $\underline{f}_1,\dots,\underline{f}_n$, $\underline{g}_1,\dots,\underline{g}_n$ such that in V the following identities hold: $\underline{p}_1(\underline{f}_1(x),x,y) = x$, $\underline{p}_i(\underline{g}_i(x),x,y) = \underline{p}_{i+1}(\underline{f}_{i+1}(x),x,y)$, $(i=1,\dots,n-1)$, $\underline{p}_n(\underline{g}_n(x),x,y) = y$.

Malcev-type-conditions are considered in (Taylor[1]) and in (Fried/Kiss[1]).

1.4 Categorical algebras

For congruence distributive varieties $V_{\underline{A}}$ we are able to give a simpler structural characterization than the characterization $V_{\underline{A}} = HSP\ \underline{A}$. This is the content of B.Jonsson's Lemma (Jonsson[1]). If a variety is generated by a finite algebra, then we obtain:

Theorem 6: If the finite non-trivial algebra $\underline{A} = \langle A,F\rangle$ generates a congruence distributive variety $V_{\underline{A}}$, then every algebra of $V_{\underline{A}}$ is isomorphic to a subdirect product of homomorphic images of subalgebras of $\underline{A}$: $V_{\underline{A}} = IP_s HS\ \underline{A}$.

The following proposition is a consequence of Theorem 6:

Corollary 7: If $\underline{A} = \langle A,F\rangle$ is a non-trivial finite simple algebra having no proper subalgebra or exactly one one-element subalgebra and $V_{\underline{A}}$ is a congruence distributive variety, then $V_{\underline{A}} = IP_s\ \underline{A}$, equivalently, every algebra of $V_{\underline{A}}$ is isomorphic to a subdirect power of $\underline{A}$.

A finite non-trivial algebra $\underline{A} = \langle A,F\rangle$ is said to be categorical if each algebra of the variety generated by $\underline{A}$ is isomorphic to a subdirect power of $\underline{A}$.

Theorem 8 (Foster[51]): Let $\underline{A} = \langle A,F\rangle$ be a categorical algebra where F contains at least one non unary function. Then it holds

(i) $\underline{A}$ is simple,

(ii) $\underline{A}$ possesses no proper subalgebra or precisely one subalgebra and this is a one-element subalgebra, and

(iii) $\underline{A}$ is equationally maximal.

Let us consider the following example. Let $\underline{A}_2 = \langle \{0,1\}, \wedge, \vee, 0\rangle$ be the two-element distributive lattice with zero-element in which

$0\wedge 0 = 0\wedge 1 = 1\wedge 0 = 0,\ 0\vee 1 = 1\vee 0 = 1\vee 1 = 1,\ 1\wedge 1 = 1,\ 0\vee 0 = 0.$

A system of axioms (see also 1.6) is given by

(1) $x \wedge (y \wedge z) = (x \wedge y) \wedge z$,

(2) $x \wedge y = y \wedge x$,

(3) $x \wedge x = x$,

(4) $x \wedge (x \vee y) = x$,

(5) $x \wedge (y \vee z) = (x \wedge y) \vee (x \wedge z)$,

(6) $x \wedge 0 = 0$,

and the dual axioms (1') - (5') which are obtained from (1) -(5) permuting $\vee$ and $\wedge$ and

(6') $x \vee 0 = x$.

Then $V_{\underline{A}_2}$ is the variety of distributive lattices with zero-element. The ternary term function d given by $d(x,y,z) = (x \wedge y) \vee (x \wedge z) \vee (y \wedge z)$ satisfies the identities $d(x,x,y) = d(x,y,x) = x$, $d(x,y,y) = y$. Hence the presumptions of Corollary 3 are satisfied and $V_{\underline{A}_2}$ is a congruence distributive variety. As a two-element algebra, $\underline{A}_2$ is simple, moreover, $\underline{A}_2$ has precisely one one-element subalgebra. Hence the presumptions of Corollary 7 are satisfied and $\underline{A}_2$ is categorical, equivalently, each distributive lattice with zero-element is isomorphic to a subdirect power of $\underline{A}_2$.

Another example is the two-element Boolean algebra $\underline{2}_B' = \langle \{0,1\}, \wedge, \vee, N \rangle$. From the preceding discussion we find that $V_{\underline{2}_B}$ is a congruence distributive variety. $\underline{2}_B$ has no proper subalgebras. Therefore $\underline{2}_B$ is categorical.

1.5 Paraprimal and cryptoprimal algebras

In 1.3-Definition 8, paraprimal algebras were defined as simple algebras generating congruence permutable varieties and containing simple subalgebras only. A cryptoprimal algebra is simple, has no proper subalgebras, and generates a congruence distributive variety. In this section some properties of paraprimal and cryptoprimal algebras will be presented.

Paraprimal algebras were introduced and studied by D.M.Clark and P.H.Krauss in [1] and [2] . They were originally defined in terms of a rather complicated combinatorial property. Clark and Krauss showed that a paraprimal algebra may also be characterized as a finite algebra $\underline{A}$ where $V_{\underline{A}}$ is congruence permutable and each non-trivial subalgebra of $\underline{A}$ is simple. We give some results which are presented in Clark/Krauss [1] , [2] .

1. An algebra $\underline{A} = \langle A,F \rangle$ satisfies the „subalgebra-condition" if every subalgebra of $\underline{A}$ has a one-element subalgebra. For paraprimal algebras with the subalgebra-condition it holds $V_{\underline{A}} = IPS\underline{A}$, hence $V_{\underline{A}} = IP_sS\underline{A}$ (by 2.4 $\underline{A}$ is said to be semicategorical).

2. An algebra $\underline{A}$ has essentially no subalgebra if all proper subalgebras of $\underline{A}$ are trivial. An algebra $\underline{A}$ is called plain if $\underline{A}$ is simple and has essentially no subalgebra. Let $\underline{A}$ be a plain and paraprimal algebra, then the following conditions are equivalent:

(i) $V_{\underline{A}} = IP_sS\underline{A}$,

(ii) $\underline{A}$ is quasiprimal or has a trivial subalgebra,

(iii) $\underline{A}$ is equationally maximal.

3. Let $\underline{A}$ and $\underline{B}$ be algebras of the same type, let φ be a congruence of $\underline{A}$ and let Ψ be a congruence of $\underline{B}$. Then $\varphi \times \Psi$ is the congruence of $\underline{A} \times \underline{B}$ where $((a_1,b_1),(a_2,b_2)) \in \varphi \times \Psi$ if and only if $(a_1,a_2) \in \varphi$ and $(b_1,b_2) \in \Psi$. Congruences of $\underline{A} \times \underline{B}$ which are not of the form $\varphi \times \Psi$ are called skew congruences of $\underline{A} \times \underline{B}$. Let $\underline{A}$ be a plain paraprimal algebra. Define

$\underline{A}^* = \underline{A}$ if $\underline{A}$ is quasiprimal. In the opposite case $\underline{A}^2$ has a skew congruence and the diagonal of $\underline{A}^2$ is a congruence class of exactly one maximal congruence of $\underline{A}^2$. In the second case we have $\underline{A}^* = \underline{A}^2/\theta$ (θ is the maximal congruence). Then $\underline{A}$ is paraprimal, moreover card A^* = card A. $\underline{A}^*$ is quasiprimal or has a one-element subalgebra. $V_{\underline{A}^*}$ is the only equationally maximal subvariety of $V_{\underline{A}}$. Then we obtain the following

Theorem 9: Let $\underline{A}$ be a plain paraprimal algebra and let $\underline{A}^*$ be the above defined algebra, then each algebra of $V_{\underline{A}}$ is isomorphic to a subdirect product of $\underline{A}^*$ and $\underline{A}$: $V_{\underline{A}} = IP_s\{\underline{A},\underline{A}^*\}$.

4. If $\underline{A}$ is plain and paraprimal, the following is equivalent:

(i) $\underline{A}^*$ is isomorphic to $\underline{A}$,

(ii) $V_{\underline{A}} = V_{\underline{A}^*}$,

(iii) $\underline{A}$ is quasiprimal or $\underline{A}$ has a trivial subalgebra.

Applying Corollary 7 we immediately obtain for cryptoprimal algebras:

Corollary 10: Each cryptoprimal algebra is categorical.

Quasiprimal and hemiprimal algebras are classes of algebras which can be defined by preservation properties of term functions (see 1.2). E.Kiss ([1]) has shown that the class of paraprimal algebras on A (card $A \geq 2$) can be defined by preservation properties of term functions if and only if A has two elements. In this case the class of paraprimal algebras can be defined by using of finitely many relations.

1.6 Finite axiomatizability of finite algebras

Let $\underline{A} = \langle A,F \rangle$ be a finite non-trivial algebra. Suppose that $\sum \subseteq \mathrm{Id}\,\underline{A}$ is a such subsystem that all identities of Id $\underline{A}$ can be derived from $\sum$ by the following consequence rules

(i) reflexivity and symmetry of the equality,

(ii) substitution of a term or of a variable for a variable in any identity,

(iii) if $\underline{p} = \underline{q}$, substitution of $\underline{p}$ for $\underline{q}$ in all places where $\underline{q}$ appears in any identity of $\sum$.

Then $\sum$ is called an axiom system of $\underline{A}$.

$\underline{A}$ is said to be finitely axiomatizable if $\underline{A}$ has a finite complete set of axioms. Many familiar algebras are finitely axiomatizable- groups, Boolean algebras, rings, lattices etc. In [1] R.C. Lyndon posed the following question: „Does every finite algebra possess a finite set of identities from which all others are derivable ?" This problem is not entirely trivial even for two-element algebras. In (Lyndon[1]) it was proved that every two-element algebra is finitely axiomatizable. We emphasize that a finite algebra has not only a finite carrier, but also a finite set of fundamental operations, i.e. a finite type. In [2] Lyndon gave an example of a seven-element algebra which is not finitely axiomatizable. Murskij proved that the three-element algebra $\underline{A} = \langle \{0,1,2\}, \cdot \rangle$, where $\cdot$ is defined by

$\cdot$	0	1	2
0	0	0	0
1	0	0	1
2	0	2	2

is not finitely axiomatizable (Murskij[1]). Under which conditions is a finite algebra finitely axiomatizable ? Here we list some classes of algebras which are finitely axiomatizable: any finite algebra which generates a congruence distributive variety (Baker[2]), any paraprimal algebra (Mc Kenzie[3]), any finite algebra whose fundamental operations essentially depend on at most one variable (Janov[1]).

2. Arithmetical varieties

2.1 Characterization of quasiprimal and functionally complete algebras

If S is a non-void set, the function $t: S^3 \to S$, defined by $t(x,y,z) = z$ if $x = y$, $t(x,y,z) = x$ if $x \neq y$, is called the ternary discriminator of S. By using the ternary discriminator quasiprimal and functionally complete algebras can be characterized in the following manner:

Theorem 1 (Pixley [3], [5]), (Werner [1]): For a finite non-trivial algebra $\underline{A} = \langle A,F \rangle$ the conditions 1), 2), 3) are equivalent and the conditions 1'), 2'), 3') are equivalent.

1) $\underline{A}$ is quasiprimal,
2) the ternary discriminator of A is a term function of $\underline{A}$,
3) each subalgebra of $\underline{A}$ is simple and for some term function q $q(x,x,z) = q(z,x,x) = q(z,x,z) = z$ are identities of $\underline{A}$.

1') $\underline{A}$ is functionally complete,
2') the ternary discriminator of A is a polynomial function of $\underline{A}$,
3') $\underline{A}$ is simple and for some polynomial function q $q(x,x,z) = q(z,x,x) = q(z,x,z) = z$ are identities of $\underline{A}$.

Proof 1)$\Rightarrow$2): Since the ternary discriminator t of A is a function preserving all subalgebras and all isomorphisms between non-trivial subalgebras of $\underline{A}$, t is a term function of $\underline{A}$.

2)$\Rightarrow$3): Let $\Theta \neq O$ (O = zero-congruence) be a congruence relation of a non-trivial subalgebra $\underline{B}$ of $\underline{A}$. Then there are $a,b \in \underline{B}$ such that $a \neq b$ and $(a,b) \in \Theta$. For any element $c \in B$ it holds $(t(a,b,c),t(a,a,c)) \in \Theta$ and hence $(a,c) \in \Theta$, i.e. $\Theta = I$ (I = universal congruence). The ternary discriminator satisfies the identities $t(x,x,z) = t(z,x,x) = t(z,x,z) = z$, therefore for q we can choose t.

3)$\Rightarrow$1) (Pixley [2]): For any natural number $n > 0$ we consider the set of all n-tuples $\alpha_i = (a_{i1},\dots,a_{in}),(i=1,\dots,k^n)$, of elements of $A = \{0,\dots,k-1\}$, $(k>1)$. Let $[\alpha_i]$ be the subalgebra of $\underline{A}$ generated by α_i. We define a partition of the α_i into equivalence classes by placing α_i and α_j in the same class provided there is an isomorphism $\varphi: \alpha_i \longrightarrow \alpha_j$ for which $\varphi(a_{i1}) = a_{j1},\dots,\varphi(a_{in}) = a_{jn}$. If $f: A^n \longrightarrow A$ is an n-ary function preserving subalgebras and isomorphisms between non-trivial subalgebras of $\underline{A}$, the value $f(\alpha_i) = f(a_{i1},\dots,a_{in})$ of f on any α_i lies in $[\alpha_i]$ and completely determines the value $f(\alpha_j)$ of f on any α_j in the same equivalence class, $(\varphi(f(\alpha_i)) = f(\alpha_j))$. If $\underline{t}(y_1,\dots,y_n)$ is a term in which n or fewer variables occur and $\underline{t}(\alpha_i) = f(\alpha_i)$ for one representative α_i from each equivalence class, then $\underline{t}$ and f agree on all α_i. Suppose that there are $m > 0$ equivalence classes of α_i and that $\alpha_1,\dots,\alpha_m$ is a complete system of representatives. Now, let f be a function preserving subalgebras and isomorphisms between non-trivial subalgebras of $\underline{A}$. To complete the proof we have to show that there is a term $\underline{t}$ such that $\underline{t}$ and f agree on $\alpha_1,\dots,\alpha_m$. Let $\underline{F}_n = \underline{F}_n(A)$ be the $V_{\underline{A}}$-free algebra freely generated by a set $G = \{y_1,\dots,y_n\}$ of n generators. Let $p_1,\dots,p_m$ be the functions from G onto the α_i such that $p_i(y_r) = a_{ir},(r=1,\dots,n)$. Each p_i can be extended to a homomorphism h_i of $\underline{F}_n$ onto $[\alpha_i]$ defined by $h_i(\underline{t}(y_1,\dots,y_n)) = \underline{t}(\alpha_i)$. Let Θ_i be the kernel of h_i, then $\underline{F}_n/\Theta_i$ is isomorphic to $[\alpha_i]$ and Θ_i puts two terms $\underline{t}_1$ and $\underline{t}_2$ in the same equivalence class if $\underline{t}_1(p_i(y_1),\dots,p_i(y_n)) = \underline{t}_2(p_i(y_1),\dots,p_i(y_n))$. Thus $(\underline{t}_1,\underline{t}_2) \in \Theta_1 \wedge \dots \wedge \Theta_m$ if and only if $\underline{t}_1(p_i(y_1),\dots,p_i(y_n)) = \underline{t}_2(p_i(y_1),\dots,p_i(y_n))$ for all $i=1,\dots,m$, $\underline{t}_1$, $\underline{t}_2 \in \underline{F}_n$. Consequently $(\underline{t}_1,\underline{t}_2) \in \Theta_1 \wedge \dots \wedge \Theta_m$ is equivalent to $\underline{t}_1 = \underline{t}_2 \in \mathrm{Id}\,\underline{A}$ i.e. equivalent to $\underline{t}_1 = \underline{t}_2$ in $\underline{F}_n$. Therefore $\Theta_1 \wedge \dots \wedge \Theta_m = 0$ and $\underline{F}_n$ is isomorphic to a subdirect product $[\alpha_1] \underline{x} \dots \underline{x} [\alpha_m]$ under the function $\Phi(\underline{t}) = (\underline{t}(\alpha_1),\dots,\underline{t}(\alpha_m))$, (by a theorem of Birkhoff [2]).

Consider a pair Θ_i,Θ_j, $i \neq j$. Since all non-trivial $[\alpha_i]$ are simple, Θ_i and Θ_j are either $= I$ or maximal. Hence $\Theta_i \vee \Theta_j = I$ or one of Θ_i,Θ_j. If $\Theta_i \vee \Theta_j = \Theta_i \neq I$, then $\Theta_i = \Theta_j$ and $[\alpha_i]$ and $[\alpha_j]$ would be isomorphic. But this is impossible since α_i, α_j are not equivalent. Therefore $\theta_i \vee \theta_j = I$ for all $i \neq j$. Congruence distributivity implies $(\theta_1 \wedge \ldots \wedge \theta_{i-1}) \vee \theta_i = (\theta_1 \vee \theta_i) \wedge \ldots \wedge (\theta_{i-1} \vee \theta_i) = I$ for $i=2,\ldots,m$. By a theorem of Birkhoff ([2]), congruence permutability, $\theta_1 \wedge \ldots \wedge \theta_m = 0$ and $(\theta_1 \wedge \ldots \wedge \theta_{i-1}) \vee \theta_i = I$ for $i=2,\ldots,m$ imply that $\underline{F}_n$ is isomorphic to a direct product of the $\underline{F}_n/\theta_i \cong [\alpha_i]$: $\underline{F}_n \cong [\alpha_1] \times \ldots \times [\alpha_m]$, i.e. Φ is an isomorphism. The term $\underline{t} = \Phi^{-1}(f(\alpha_1),\ldots,f(\alpha_m))$ agrees with f on $\alpha_1,\ldots,\alpha_m$ and, consequently, on all α_i as shown above.

<u>1')⟹2'):</u> All functions on A, in particular the ternary discriminator, are polynomial functions of $\underline{A}$.

<u>2')⟹3'):</u> Let $\underline{A}^+$ be the constantive algebra obtained from $\underline{A}$ by adding new constant term functions, one for each element of A. If the ternary discriminator is a polynomial function of $\underline{A}$, then it is a term function of $\underline{A}^+$. Therefore $\underline{A}^+$ is simple (by 2)⟹3)). $\underline{A}$ and $\underline{A}^+$ have the same congruence relations, hence $\underline{A}$ is also simple. Since the ternary discriminator, as a polynomial function, satisfies the identities $t(x,x,z) = t(z,x,x) = t(z,x,z) = z$, the second part of 3') is clear, too.

<u>3')⟹1'):</u> Since $\underline{A}$ is simple, $\underline{A}^+$ is also simple. $\underline{A}^+$ has, as a constantive algebra, no proper subalgebras and no non-identical automorphisms, (see 1.2). The function q with $q(x,x,z) = q(z,x,x) = q(z,x,z) = z$ is a polynomial function of $\underline{A}$, therefore q is a term function of $\underline{A}^+$. From the first part of the proof it follows that $\underline{A}^+$ is quasiprimal. $\underline{A}^+$ has no proper subalgebras and no non-identical automorphisms, consequently, all functions on A are term functions of $\underline{A}^+$, i.e. $\underline{A}^+$ is primal and $\underline{A}$ is functionally complete. ▲

As a consequence of Theorem 1 we obtain:

<u>Corollary 2 (Werner [1]):</u> Each quasiprimal algebra is functionally complete.

From the condition 3) of Theorem 1 and the Malcev-type -condition 1.3-Corollary 3 it follows that quasiprimal algebras generate arithmetical varieties. Clearly, primal, demiprimal, and semiprimal algebras generate arithmetical varieties, too.

Under which additional conditions is an algebra primal, demiprimal, semiprimal or quasiprimal provided that it generates an arithmetical variety ?

We obtain the following proposition:

A non-trivial finite algebra is quasiprimal if and only if

(i) $V_{\underline{A}}$ is arithmetical,

(ii) each non-trivial subalgebra of $\underline{A}$ is simple.

2.2 A characterization of semiprimal, demiprimal, and primal algebras

Theorem 3: A finite non-trivial algebra $\underline{A} = \langle A,F\rangle$ is semiprimal iff

(i) $V_{\underline{A}}$ is arithmetical,

(ii) all non-trivial subalgebras of $\underline{A}$ are simple,

(iii) all subalgebras of $\underline{A}$ are rigid, (i.e. all subalgebras of $\underline{A}$ have no non-identical automorphisms),

(iv) no two distinct subalgebras of more than one element are isomorphic.

Proof: If $\underline{A}$ is semiprimal, $\underline{A}$ is also quasiprimal. Consequently, $V_{\underline{A}}$ is arithmetical and all non-trivial subalgebras of $\underline{A}$ are simple. From the definition of semiprimal algebras it follows that any non-trivial subalgebra of a semiprimal algebra is semiprimal. Let φ be a non-identical automorphism of $\underline{A}$, $a,b \in A$, $\varphi(a) = b \neq a$, $\varphi(b) = b'$. Then $b \neq b'$. Consider the function f with $f(x,y) = x$ if $x \neq a$, $y \neq b$ and $f(a,b) = b$. This function preserves all subalgebras of $\underline{A}$ and therefore f is a term function of $\underline{A}$. Then $f(\varphi(a),\varphi(b)) = f(b,b') = b$, $\varphi(f(a,b)) = \varphi(b) = b'$. This contradicts our assumption that φ is an automorphism. Since every non-trivial subalgebra of $\underline{A}$ is semiprimal, all subalgebras of $\underline{A}$ are rigid. To demonstrate (iv) let $\underline{S}_1$ and $\underline{S}_2$ be two distinct isomorphic subalgebras of $\underline{A}$. Since $\underline{A}$ is finite, no proper subalgebra of $\underline{S}_2$ can be isomorphic to $\underline{S}_2$ and hence we may suppose that the set $S_1 - S_2$ is non-empty. Let $a \in S_1 - S_2$ and let $b \in S_2$ be the isomorphic image of a: $\varphi(a) = b$. Further, let $c \neq a$ be any other element of S_1 and $\varphi(a) = d \in S_2$, $d \neq b$. The function f with $f(x,y) = x$ if $x = a$ and $f(x,y) = y$ otherwise, is a term function of $\underline{A}$. Under the supposed isomorphism we then have $f(\varphi(a),\varphi(c)) = f(b,d) = d$, $\varphi(f(a,c)) = \varphi(a) = b$, which is a contradiction. This completes the „only if" part of the proof.

Now we will prove that from (i)-(iv) it follows the semiprimality of $\underline{A}$. Applying Theorem 1 from (i) and (ii) it follows that $\underline{A}$ is quasiprimal. There are no isomorphisms between non-trivial subalgebras of $\underline{A}$, therefore each function preserving all subalgebras of $\underline{A}$ is a term function of

$\underline{A}$. Consequently, $\underline{A}$ is semiprimal. This completes the proof. ▲

Clearly, a semiprimal algebra with no proper subalgebra is primal. A semiprimal algebra $\underline{A}$ which is not primal has at least one minimal subalgebra, i.e. one having no proper subalgebras. Each minimal subalgebra of a semiprimal algebra is either primal or, else, is a one-element subalgebra. If $\underline{A}$ has more than one minimal subalgebra, any two such subalgebras are disjoint. By a subprimal algebra we mean a semiprimal $\underline{A}$ with precisely one subalgebra $\underline{A}^*$ ($\neq \underline{A}$), called the core of $\underline{A}$. $\underline{A}$ is called regular if $\underline{A}^*$ contains at least two elements and singular otherwise.

In [6] Foster defined the following notions: An element $a \in A$ is called a centroid element of $\underline{A} = \langle A,F \rangle$ if there exists a unary term function c_a of $\underline{A}$ with $c_a(x) \equiv a$ for all $x \in A$. The set $C(\underline{A})$ of all centroid elements of $\underline{A}$ is called the centroid of $\underline{A}$.

<u>Lemma 4:</u> The centroid $C(\underline{A})$ of $\underline{A}$, if not empty, is a subalgebra of $\underline{A}$ which is a subalgebra of every subalgebra of $\underline{A}$. ($C(\underline{A})$ is the least subalgebra).

<u>Proof:</u> Let $a_1,\ldots,a_n \in C(\underline{A})$, i.e. there exist term functions $c_1,\ldots,c_n$ of $\underline{A}$ such that $c_1(x) \equiv a_1,\ldots,c_n(x) \equiv a_n$. We have $f(c_1(x),\ldots,c_n(x)) = b$ for any fundamental operation f of $\underline{A}$ where $b = f(a_1,\ldots,a_n)$. Since $f(c_1(x),\ldots,c_n(x))$ is a term function of $\underline{A}$, it follows $b \in C(\underline{A})$. Let $\underline{B}$ be a subalgebra of $\underline{A}$ and let $C(\underline{A})$ be the centroid of $\underline{A}$. For any $b \in C(\underline{A})$ there exists a term function b with $c_b(x) \equiv b$ for all $x \in A$, in particular for all $x \in B \subseteq A$. c_b is a term function of $\underline{B}$ thus $b \in B$, i.e. $C(\underline{A}) \subseteq \underline{B}$. ▲

Obviously, if the centroid $C(\underline{A})$ is not empty, it holds $C(\underline{A}) = \underline{A}_1 \cap \ldots \cap \underline{A}_m$ where $\underline{A}_1,\ldots,\underline{A}_m$ are all subalgebras of $\underline{A}$.

An algebra $\underline{A} = \langle A,F \rangle$ is constantive (see 1.2) if and only if $C(\underline{A}) = \underline{A}$.

Corollary 5 (Foster/Pixley[1],[2]): If $\underline{A}$ is a subprimal algebra, its centroid $C(\underline{A})$ is non-empty and is identical with its core, $C(\underline{A}) = \underline{A}^*$. The centroid of $\underline{A}$ is a primal algebra or else a one-element algebra.

Now we consider demiprimal algebras.

Theorem 6: A finite non-trivial algebra $\underline{A} = \langle A,F \rangle$ is demiprimal iff

(i) $V_{\underline{A}}$ is arithmetical,

(ii) $\underline{A}$ has no proper subalgebras,

(iii) $\underline{A}$ is simple,

(iv) all non-identical automorphisms of $\underline{A}$ are representable by permutations of $\underline{A}$ without fixed elements.

Proof: If $\underline{A}$ is demiprimal, $\underline{A}$ is also quasiprimal. Therefore, by Theorem 1, $\underline{A}$ is simple and $V_{\underline{A}}$ arithmetical. Conversely, applying Theorem 1 from (i),(ii),(iii) it follows that $\underline{A}$ is quasiprimal. Since $\underline{A}$ has no proper subalgebras, each isomorphism between non-trivial subalgebras of $\underline{A}$ is an automorphism and $\underline{A}$ is demiprimal. We prove that all non-identical automorphisms of a demiprimal algebra $\underline{A}$ are representable by permutations of A without fixed elements. Any automorphism of $\underline{A}$ is representable by a permutation s of A. Let S be the set of all fixed elements of s and $S \neq \emptyset$. Then $\underline{S}$ is a subalgebra of $\underline{A}$. A demiprimal algebra has no proper subalgebra, therefore $\underline{S} = \underline{A}$ and s is the identical automorphism. ▲

Applying Theorem 3 and Theorem 6 we obtain:

Corollary 7: A non-trivial finite algebra $\underline{A} = \langle A,F \rangle$ is primal iff

(i) $V_{\underline{A}}$ is arithmetical,

(ii) $\underline{A}$ has no proper subalgebras,

(iii) $\underline{A}$ is rigid (has no non-identical automorphisms),

(iv) $\underline{A}$ is simple.

Proof: Applying Theorem 3 it follows that $\underline{A}$ is semiprimal. Semiprimal algebras with no proper subalgebras are primal. If $\underline{A}$ is primal, then $\underline{A}$ is also semiprimal and demiprimal. Thus (i),(ii),(iii),(iv) are valid. ▲

2.3 Hemiprimal algebras

Now we shall characterize arithmetical hemiprimal algebras. The following lemma is well-known for the case of rings (Zariski/Samuel [1]). It was proved for arbitrary algebras in (Sioson [1]).

Lemma 8: An algebra $\underline{A}$ is arithmetical if and only if for each finite set $\theta_1, \ldots, \theta_n$ of congruences of $\underline{A}$ and elements $x_1, \ldots, x_n \in A$ for which $(x_i, x_j) \in \theta_i \cup \theta_j$, $(i,j=1,\ldots,n)$ there is an $x \in A$ for $(x, x_i) \in \theta_i$.

Lemma 9: If $\underline{A}$ is an arbitrary non-trivial algebra whose congruence lattice is finite, $\underline{A}$ is arithmetical if and only if there is a function $f: A^3 \to A$ which preserves all congruences of $\underline{A}$ and which satisfies $f(x,x,z) = f(z,x,x) = f(z,x,z) = z$ for all $x,z \in A$.

Lemma 9 was proved in Pixley [5]. Note that the identities have the same form as the Malcev-type-condition characterizing arithmetical varieties (see 1.3).

We get:

Theorem 10 (Pixley [5]): Let $\underline{A} = \langle A,F\rangle$ be a finite non-trivial arithmetical algebra, then $\underline{A}$ is hemiprimal if and only if

(i) $V_{\underline{A}}$ is arithmetical,

(ii) $\underline{A}$ has no proper subalgebras,

(iii) if θ_1, θ_2 are congruences of $\underline{A}$, then any isomorphism $\varphi: \underline{A}/\theta_1 \to \underline{A}/\theta_2$ is the identity, and, in particular, $\theta_1 = \theta_2$.

Proof: Suppose $\underline{A}$ is hemiprimal. Applying Lemma 9, there is a term function f of $\underline{A}$ with $f(x,x,z) = f(z,x,x) = f(z,x,z) = z$ for all $x,z \in A$. Thus the Malcev-type-condition characterizing arithmetical varieties is satisfied and $V_{\underline{A}}$ is arithmetical. Since all constant functions are term functions of $\underline{A}$, $\underline{A}$ is constantive, i.e. $C(\underline{A}) = \underline{A}$, thus $\underline{A}$ has no proper subalgebras. If $\varphi: \underline{A}/\theta_1 \to \underline{A}/\theta_2$ is an isomorphism and $a \in A$, then consider the constant term function c_a with $c_a(x) = a$ for all $x \in A$. Then

$\varphi([a]_{\theta_1}) = (c_a([x]_{\theta_1})) = c_a(\varphi([x]_{\theta_1})) = [a]_{\theta_2}$. Hence $[a]_{\theta_1} = [b]_{\theta_1}$ implies $[a]_{\theta_2} = [b]_{\theta_2}$ and conversely. Thus $\theta_1 = \theta_2$ what proves (iii).

Conversely, suppose (i)-(iii) hold for $\underline{A}$. For a natural number n let $f: A^n \longrightarrow A$ be a function which preserves all congruences of $\underline{A}$. Let $\underline{F}_n = \underline{F}_n(\underline{A})$ be the free algebra freely generated by the set $G = \{y_1,\dots,y_n\}$. Let p_i, $(i=1,\dots,k^n)$, be all functions of G into A. Let h_i be the unique homomorphism of $\underline{F}_n$ into $\underline{A}$ extending p_i. h_i is „onto" by (ii). Let Φ_i be the kernel of h_i and for each i choose a term $\underline{t}_i(y_1,\dots,y_n)$ in $\underline{F}_n$ such that $h_i(\underline{t}_i(y_1,\dots,y_n)) = f(p_i(y_1),\dots,p_i(y_n))$. For each $i,j = 1,\dots,k^n$, there are congruences θ_i, θ_j on $\underline{A}$ and isomorphisms φ_i, φ_j, $\varphi_i: \underline{F}_n/(\Phi_i \cup \Phi_j) \longrightarrow \underline{A}/\theta_i$, $\varphi_j: \underline{F}_n/(\Phi_i \cup \Phi_j) \longrightarrow \underline{A}/\theta_j$ where, in particular, for $m=1,\dots,n$ $\varphi_i([y_m]_{\Phi_i \cup \Phi_j}) = [p_i(y_m)]_{\theta_i}$ and $\varphi_j([y_m]_{\Phi_i \cup \Phi_j}) = [p_j(y_m)]_{\theta_j}$. But $\varphi_j \circ \varphi_i^{-1}: \underline{A}/\theta_i \longrightarrow \underline{A}/\theta_j$ is an isomorphism, therefore $\varphi_i = \varphi_j$ and $\theta_i = \theta_j$ $(=\theta)$, (by (iii)). Consequently, $(p_i(y_m), p_j(y_m)) \in \theta$, $m=1,\dots,n$, and hence
$(f(p_i(y_1),\dots,p_i(y_n)), f(p_j(y_1),\dots,p_j(y_n))) \in \theta$
since f preserves all congruences. But this means that
$(h_i(\underline{t}_i(y_1,\dots,y_n)), h_j(\underline{t}_j(y_1,\dots,y_n))) \in \theta$ thus $(\underline{t}_i(y_1,\dots,y_n), \underline{t}_j(y_1,\dots,y_n)) \in (\Phi_i \cup \Phi_j)$ for all $i,j = 1,\dots,k^n$. By (i) $\underline{F}_n$ is arithmetical and thus by Lemma 8 there is a $\underline{t}(y_1,\dots,y_n) \in \underline{F}_n$ such that
$(\underline{t}(y_1,\dots,y_n), \underline{t}_i(y_1,\dots,y_n)) \in \Phi_i$, $i=1,\dots,k^n$.
This implies
$\underline{t}(p_i(y_1),\dots,p_i(y_n)) = f(p_i(y_1),\dots,p_i(y_n))$ for $i=1,\dots,k^n$ and hence that $\underline{t}$ and f agree on $\underline{A}$. Thus $\underline{A}$ is hemiprimal. ▲

The structure theory for hemiprimal algebras was investigated by Foster and Pixley (Foster[8], Pixley[4]).

A hemiprimal algebra $\underline{A}$ is called linear hemiprimal if the congruence lattice of $\underline{A}$ is a chain. Linear hemiprimal algebras are arithmetical and thus Theorem 10 can be applied.

2.4 Representation theorem

From the results of 2.2 and 2.3 we obtain the following representation theorem for semi-, demi-, and hemiprimal algebras applying 1.4-Theorem6.

<u>Theorem 11:</u> Let $\underline{A} = \langle A,F \rangle$ be a finite non-trivial algebra and let $V_{\underline{A}}$ be the variety generated by $\underline{A}$. Then it holds:

1. If $\underline{A}$ is semiprimal, then $V_{\underline{A}} = IP_sS\underline{A}$ (we say that $\underline{A}$ is semicategorical). If $\underline{A}$ is singular subprimal, then $V_{\underline{A}} = IP_s\underline{A}$, i.e. $\underline{A}$ is categorical.
2. If $\underline{A}$ is hemiprimal, then $V_{\underline{A}} = IP_sH\underline{A}$ (in this case $\underline{A}$ is said to be hemicategorical).
3. If $\underline{A}$ is demiprimal, then $V_{\underline{A}} = IP_s\underline{A}$, i.e. $\underline{A}$ is categorical.

In particular, primal algebras are categorical.

Problems and comments

1. Find a set of fundamental operations for every three-element semiprimal algebra.
2. Let p be a prime number and $\underline{H_p} = \langle \{0,\dots,p-1\},+,*\rangle$ (Foster/Pixley[2]) be the algebra of p elements in which x + y is the ordinary addition mod p and $x * y = \min(x,y)$ in the ordering $0,2,3,\dots,p-1,1$. Prove that $\underline{H_p}$ is a singular subprimal algebra.
3. Give an example of an algebra $\underline{A}$ with a subalgebra $\underline{B}$ such that $C(\underline{B}) \neq \emptyset$, but $C(\underline{A}) = \emptyset$.
4. Is every functionally complete algebra finitely axiomatizable ?

3. Two-element preprimal algebras

The two-element Boolean algebra generates the variety of Boolean algebras. It is primal and therefore, by 2.4, it is categorical. The importance of the two-element Boolean algebra as the algebra of the truth-values of the classical propositional calculus has influenced the study of other two-element algebras. In various branches of Universal algebra we find two-element algebras. Let us consider some examples:

In (Janov [1]) it is proved: Let $\underline{A} = \langle A,F \rangle$ be a finite non-trivial algebra such that any term function of $\underline{A}$ depends on all its arguments, then there is a two-element algebra $\underline{2}$ with Id $\underline{A} \subseteq$ Id $\underline{2}$, i.e. $V_{\underline{2}} \subseteq V_{\underline{A}}$. In (Taylor [2]) the notion of the fine spectrum $f_V(n)$ of a variety V was introduced. Hereby $f_V(n)$ denotes the number of isomorphism types of algebras of cardinality n in V. In 3.2 we shall see that varieties with certain given fine spectra are generated by two-element algebras. Another notion where two-element algebras occur, is the notion of an independent set of an algebra in the sense of Marzewski ([1]). In fact, each non-trivial algebra in which all elements are independent, is a two-element algebra.

There are also many varieties which are generated by two-element algebras, e.g. distributive lattices, Boolean rings, Boolean 3-rings, Boolean 3-algebras, Boolean groups, Boolean 3-groups (Frink [1], Grau [1], [2], Kempe [1], Monk/Sioson [1], Post [2], Whiteman [1]).

But also for applications, e.g. for synthesis of switching circuits, two-element algebras play an important role. Moreover, the result of Lyndon ([1]), that every two-element algebra is finitely axiomatizable, is essential for computer science.

In this chapter we shall investigate two-element algebras which are not so far from Boolean algebras, more concrete, which are preprimal. Section 3.2 deals with all two-element algebras and the varieties generated by these algebras.

3.1 Characterization of two-element preprimal algebras

Consider the two-element set $\{0,1\}$ and let $f:\{0,1\}^n \longrightarrow \{0,1\}$ be an n-ary function on $\{0,1\}$. A function $f \in \underline{F}(\{0,1\}) = \bigcup_{n=0}^{\infty} F_n(\{0,1\})$ is called a Boolean function. For Boolean functions and closed classes of Boolean functions, (i.e. closed under superposition), we use the notations given e.g. in (Post [1], [3]) and in (Jablonskij, Gawrilow, Kudrjawzew [1]).

$0,1$ nullary Boolean functions, $Nx = 1-x$ negation

$x \vee y = \max(x,y)$ disjunction, $x \wedge y = \min(x,y)$ conjunction,

$x+y = \begin{cases} 0 \text{ if } x=y \\ 1 \text{ if } x \neq y \end{cases}$ addition mod 2, $x \equiv y = (x \supset y) \wedge (y \supset x)$ equivalence,

$x \supset y = \begin{cases} 1 \text{ if } x \leq y \\ 0 \text{ otherwise} \end{cases}$ implication, $(x,y,z) = x \wedge (y \vee z)$,

$[x,y,z] = x \wedge (y \equiv z)$,

$d_n(x_1,\ldots,x_n) = x_2x_3\ldots x_n \vee x_1x_3\ldots x_n \vee \ldots \vee x_1\ldots x_{n-2}x_{n-1}$ for all $n \geq 3$. (Simplifying, instead of d_3 we will write d and instead of $x_2 \wedge \ldots \wedge x_n$ we will write $x_2\ldots x_n$.) A Boolean function f^* is said to be dual to the Boolean function f if $f^*(x_1,\ldots,x_n) = Nf(Nx_1,\ldots,Nx_n)$ for all $x_1,\ldots,x_n \in \{0,1\}$. f is said to be self-dual if $f^* = f$. Let $\mathfrak{s}$ be the ordinary order relation on $\{0,1\}$. An n-ary Boolean function which preserves $\mathfrak{s}$ is said to be monotonic. An n-ary Boolean function f is said to be linear if there are $c_0,\ldots,c_n \in \{0,1\}$ such that $f(x_1,\ldots,x_n) = c_0+c_1x_1+\ldots+c_nx_n$ for all $x_1,\ldots,x_n$. A Boolean function f is called to be 1-separable (0-separable) if $f(a_1,\ldots,a_1,\ldots,a_n) = 1$, ($f(a_1,\ldots,a_1,\ldots,a_n) = 0$), $a_1 \in \{0,1\}$, $1=1,\ldots,n$, implies $a_i=1$, ($a_i=0$), for some $i \in \{1,\ldots,n\}$. A Boolean function is called to be 1-separable, (0-separable), of degree n, ($n \geq 2$), if every n-element subset of $f^{-1}(1)$, ($f^{-1}(0)$), is 1-separable, (0-separable). For an algebra $\underline{A} = \langle A,F \rangle$ let $\underline{A}^0 = \langle A,F^0 \rangle$ be the algebra such that F^0 contains all dual functions of F. Then $\underline{A}$ and $\underline{A}^0$ are isomorphic, (more precisely, dually isomorphic), and we make no distinction between them.

Post proved in [3] that there are exactly the following maximal classes of Boolean functions and he described a basis for each of these classes.

1. class of all Boolean functions preserving{0}, ({1}), basis: $\langle +, \wedge, 0\rangle$ $(\langle x+y+1, \vee, 1\rangle)$,
2. class of all monotonic Boolean functions, basis: $\langle \wedge, \vee, 0, 1\rangle$,
3. class of all self-dual Boolean functions, basis: $\langle d, x+y+z, N\rangle$,
4. class of all linear Boolean functions, basis: $\langle +, N, 0, 1\rangle$.

Up to equivalence we have exactly the following two-element preprimal algebras: $\underline{C}_3 = \langle\{0,1\}, +, \wedge, 0\rangle$, ($\underline{C}_2 = \langle\{0,1\}, x+y+1, \vee, 1\rangle$ is dually isomorphic to $\underline{C}_3$), $\underline{A}_1 = \langle\{0,1\}, \wedge, \vee, 0, 1\rangle$, $\underline{D}_3 = \langle\{0,1\}, d, x+y+z, N\rangle$, $\underline{L}_1 = \langle\{0,1\}, +, N, 0, 1\rangle$.

We will investigate which two-element preprimal algebras generate congruence permutable varieties and which generate congruence distributive varieties. Using 1.3- Theorem 1 and 1.3- Corollary 3 we see that $V_{\underline{C}_3}$, $V_{\underline{L}_1}$, $V_{\underline{D}_3}$ are congruence permutable, thus $\underline{C}_3$, $\underline{L}_1$, $\underline{D}_3$ are paraprimal, and that $V_{\underline{A}_1}$, $V_{\underline{C}_3}$, $V_{\underline{D}_3}$ are congruence distributive. In fact, the function p defined by $p(x,y,z) = x+y+z$ satisfies the identities of 1.3- Theorem 1. Moreover, the function d defined by $d(x,y,z) = (x \wedge y) \vee (y \wedge z) \vee (x \wedge z)$ satisfies the identities of 1.3 - Corollary 3.

$V_{\underline{A}_1}$ is not congruence permutable since a function p with $p(x,x,y) = y$, $p(x,y,y) = x$ is not monotonic. $V_{\underline{L}_1}$ is not congruence distributive since $V_{\underline{L}_1}$ is congruence permutable and a function q with $q(x,x,z) = q(z,x,x) = q(z,x,z) = z$, (1.3 - Corollary 4), is not linear.

Let us collect some more properties for preprimal algebras. $\underline{C}_3$ is semiprimal (by Definition 1) and $\underline{D}_3$ is demiprimal (by Definition 2). Applying 1.4- Corollary 7 we verify that $\underline{C}_3$, $\underline{A}_1$, and $\underline{D}_3$ are categorical, thus they are equationally maximal by 1.4- Theorem 8. $\underline{L}_1$ is plain and paraprimal, but $\underline{L}_1$ is not quasiprimal. By 1.5 there is a maximal congruence θ of $\underline{L}_1^2$ with $\underline{L}_1^* = \underline{L}_1^2/\theta$ and $V_{\underline{L}_1} = IP_s\{\underline{L}_1, \underline{L}_1^*\}$. ($\underline{L}_1^*$ is equivalent to $\underline{L}_3 = \langle\{0,1\}, +, 0\rangle$). There are exactly two subdirectly irreducible algebras in $V_{\underline{L}_1}$: $\underline{L}_1$ and $\underline{L}_1^*$. $V_{\underline{L}_1^*}$ is the only non-trivial subvariety of

$V_{\underline{L}_1}$.

Now we are ready to give an algebraic characterization of two-element preprimal algebras:

Theorem 1: Let $\underline{A}$ be a two-element non-primal algebra. Then $\underline{A}$ is preprimal if and only if one of the following cases is satisfied:

1. $\underline{A}$ is demiprimal (Definition 2),
2. $\underline{A}$ is singular subprimal (2.2),
3. $\underline{A}$ is paraprimal and constantive (Definition 8),
4. $\underline{A}$ is cryptoprimal and constantive (Definition 7).

Proof: The "only if" part of the proof is clear since $\underline{D}_3$ is demiprimal, $\underline{C}_3$ is singular subprimal, $\underline{L}_1$ is paraprimal and constantive, and $\underline{A}_1$ is cryptoprimal and constantive. Suppose one of the conditions 1.,2.,3., 4. is fulfilled. In the first case we have $\underline{A} = \underline{D}_3$ because of Definition 2. Note that all self-dual functions are the term functions of $\underline{D}_3$. If we suppose 2., then from the definitions directly follows $\underline{A} = \underline{C}_3$ or $\underline{A} = \underline{C}_2$. In the third case there is a ternary term function p with $p(x,x,y) = y$, $p(x,y,y) = x$ in $\underline{A}$. Since $\underline{A}$ is constantive, 0 and 1 are nullary term functions of $\underline{A}$. We obtain $p(0,x,1) = Nx$, $p(x,0,y) = x+y$, i.e. $\underline{A}$ is equivalent to $\underline{L}_1=\langle\{0,1\},+,N,0,1\rangle$ and thus $\underline{A}$ is preprimal. If we suppose 4., then the functions of 1.3- Theorem 2 are term functions of $\underline{A}$. Then $t_0(x,x,z) = t_1(x,x,z) = t_1(x,y,x) = x$ for all $x,y \in \{0,1\}$. If $t_1(1,0,0) = 1$ and $t_1(0,1,1) = 0$, then t_1 is a projection. Therefore $t_1(1,0,0) = 0$ or $t_1(0,1,1) = 1$. If $t_1(1,0,0) = 0$, then it holds $t_1(1,x,y) = x \vee y$ and $t_1(x,0,y) = x \wedge y$. If $t_1(0,1,1) = 1$, then we have $t_1(0,x,y) = x \wedge y$ and $t_1(x,1,y) = x \vee y$. This shows that $\underline{A}$ is equivalent to $\underline{A}_1=\langle\{0,1\},\wedge,\vee,0,1\rangle$, therefore $\underline{A}$ is preprimal.▲

By Theorem 1 every two-element preprimal algebra generates a congruence distributive or a congruence permutable variety. We note that a two-element preprimal algebra is either quasiprimal (1.,2.) or constantive (3.,4.).

3.2 Two-element algebras and varieties generated by two-element algebras

A complete system of representatives of all two-element algebras was given in (Post[3]). It should be emphasized that here two-element algebras are studied only up to equivalence. Let us list all two-element algebras with at least one non-nullary term function:

$\underline{O}_4 = \langle\{0,1\},N\rangle$, $\underline{O}_9 = \langle\{0,1\},N,0\rangle$, $\underline{S}_3 = \langle\{0,1\},\wedge,1\rangle$,

$\underline{S}_1 = \langle\{0,1\},\wedge\rangle$, $\underline{S}_4 = \langle\{0,1\},\wedge,0\rangle$, $\underline{A}_4 = \langle\{0,1\},\wedge,\vee\rangle$,

$\underline{S}_6 = \langle\{0,1\},\wedge,0,1\rangle$ $\underline{A}_1 = \langle\{0,1\},\wedge,\vee,0,1\rangle$, $\underline{C}_3 = \langle\{0,1\},+,\wedge,0\rangle$,

$\underline{A}_2 = \langle\{0,1\},\wedge,\vee,0\rangle$, $\underline{L}_1 = \langle\{0,1\},+,N,0,1\rangle$. $\underline{L}_5 = \langle\{0,1\},x+y+z,N\rangle$,

$\underline{L}_3 = \langle\{0,1\},+,0\rangle$, $\underline{L}_4 = \langle\{0,1\},x+y+z\rangle$, $\underline{F}_4 = \langle\{0,1\},\supset,1\rangle$,

$\underline{F}_4^n = \langle\{0,1\},\supset,1,d_n\rangle$ for all $n \geq 3$, $\underline{F}_6 = \langle\{0,1\},(x,y,z),xy\rangle$,

$\underline{F}_7^n = \langle\{0,1\},(x,y,z),d_n,xy,0\rangle$ for all $n \geq 3$ $\underline{F}_7 = \langle\{0,1\},(x,y,z),xy,0\rangle$,

$\underline{F}_5 = \langle\{0,1\},[x,y,z],(x,y,z),xy\rangle$, $\underline{C}_4 = \langle\{0,1\},[x,y,z],(x,y,z),xy,x\vee y\rangle$,

$\underline{F}_7^n = \langle\{0,1\},(x,y,z),d_n,xy,0\rangle$ for all $n \geq 3$,

$\underline{F}_5^n = \langle\{0,1\},[x,y,z],(x,y,z),d_n,xy\rangle$ for all $n \geq 3$,

$\underline{D}_2 = \langle\{0,1\},d\rangle$, $\underline{D}_1 = \langle\{0,1\},d,x+y+z\rangle$, $\underline{D}_3 = \langle\{0,1\},d,x+y+z,N\rangle$.

(We give one representative for each class of isomorphic algebras.)

Now we are going to investigate the properties of the congruence lattices of all algebras in a variety generated by these two-element algebras. It is easy to show that each of the algebras $\underline{L}_3$, $\underline{L}_1$, $\underline{C}_3$, $\underline{C}_1$, $\underline{L}_4$, $\underline{L}_5$, $\underline{D}_1$, $\underline{D}_3$ has $p(x,y,z) = x+y+z$ as a term function. Thus each of these algebras generates a congruence permutable variety (1.3- Theorem 1). $p(x,y,z) = [x,y,z]\vee[z,y,x]$, is a term function of $\underline{C}_4$ and p fulfills the conditions of 1.3- Theorem 1 (Denecke[1]). A result of (Denecke[1]) shows that only the algebras $\underline{L}_1$, $\underline{L}_3$, $\underline{L}_4$, $\underline{L}_5$, $\underline{C}_1$, $\underline{C}_3$, $\underline{C}_4$, $\underline{D}_1$, $\underline{D}_3$ are two-element paraprimal algebras. The congruence distributivity of $V_{\underline{A}_1}$, $V_{\underline{A}_2}$, $V_{\underline{A}_4}$, $V_{\underline{C}_1}$, $V_{\underline{C}_3}$, $V_{\underline{C}_4}$, $V_{\underline{D}_1}$, $V_{\underline{D}_2}$, $V_{\underline{D}_3}$ is a consequence of 1.3- Corollary 2 with $d(x,y,z) = (x\wedge y)\vee(y\wedge z)\vee(x\wedge z)$, $((x\vee y = (x\wedge y)+(x+y))$.

By 1.3-Theorem 2 $\underline{F}_4$ and $\underline{F}_4^n$ generate congruence distributive varieties since $t_0(x,y,z) = x$, $t_1(x,y,z) = (((x\supset y)\supset z)\supset(y\supset x))\supset x$,

$t_2(x,y,z) = (x \supset y) \supset z$, $t_3(x,y,z)$ are term functions of $\underline{F}_4$ and $\underline{F}_4^n$ and these functions satisfy the identities of Theorem 2. $\underline{F}_4$ and $\underline{F}_4^n$ are 3-distributive but they are not 2-distributive since $\langle \supset, 1 \rangle$, $(\langle \supset, d_n, 1 \rangle)$ generate the closed class of all 0-separable, (0-separable of degree n), Boolean functions, but the function d with $d(x,x,y) = d(x,y,x) = x$, $d(x,y,y) = y$ is not 0-separable, (0-separable of degree n). Since $d(0,0,1) = d(0,1,0) = d(1,0,0) = 0$, d is not a term function of $\underline{F}_4$, $(\underline{F}_4^n)$, (Denecke[3]). By 1.3-Theorem 2 $\underline{F}_5, \underline{F}_6, \underline{F}_7, \underline{F}_5^n, \underline{F}_6^n, \underline{F}_7^n$ generate congruence distributive varieties since t_0, t_1, t_2, t_3, t_4 with $t_0 = x$, $t_1(x,y,z) = (x,y,z)$, (for the definition of (x,y,z) see 3.1), $t_2(x,y,z) = xy$, $t_3(x,y,z) = (z,y,x)$, $t_4 = z$ are term functions of $\underline{F}_5$, $\underline{F}_6, \underline{F}_7, \underline{F}_5^n, \underline{F}_6^n, \underline{F}_7^n$ and the identities of 1.3-Theorem 2 are satisfied.
We remark that in (Baker[1]) these functions are used to prove the congruence distributivity of a reduct $\underline{L}_p = \langle L, (x,y,z) \rangle$ of a lattice $\underline{L} = \langle L, \wedge, \vee \rangle$ and to show that the algebras $\underline{L}_p = \langle L, (x,y,z) \rangle$ are 4-distributive but not 3-distributive. It follows that also the algebras $\underline{F}_5, \underline{F}_6, \underline{F}_7, \underline{C}_4, \underline{F}_5^n, \underline{F}_6^n, \underline{F}_7^n$ are 4-distributive and not 3-distributive.
Only the two-element algebras $\underline{A}_1, \underline{A}_2, \underline{A}_4, \underline{D}_1, \underline{D}_2, \underline{D}_3, \underline{C}_1, \underline{C}_3, \underline{C}_4, \underline{F}_4, \underline{F}_5, \underline{F}_6, \underline{F}_7$, $\underline{F}_4^n, \underline{F}_5^n, \underline{F}_6^n, \underline{F}_7^n$ generate congruence distributive varieties and there are (up to equivalence) only finitely many two-element algebras not generating congruence distributive varieties.

Applying these results on the varieties generated by two-element algebras we obtain the following consequences:

1. The two-element algebras $\underline{C}_1, \underline{C}_3, \underline{C}_4, \underline{D}_1, \underline{D}_3$ are quasiprimal, $\underline{C}_1, \underline{C}_3, \underline{C}_4$ are semiprimal, $\underline{D}_3$ is demiprimal.
2. Two-element algebras are paraprimal or they generate congruence distributive varieties or they have one of the forms $\underline{O}_4, \underline{O}_9, \underline{S}_1, \underline{S}_3, \underline{S}_4$, $\underline{S}_6$.

(From 2. we get the following propositions: $\underline{S}_1, \underline{S}_3, \underline{S}_4, \underline{S}_6$ are finitely axiomatizable since each of these algebras is a semi-lattice. $\underline{O}_4$ and $\underline{O}_9$ are finitely axiomatizable since the fundamental operations of $\underline{O}_4$ and $\underline{O}_9$

essentially depend on at most one variable (1.6,3). By 1.6,1 and 1.6,2 all other two-element algebras are finitely axiomatizable.)

From the congruence distributivity and the congruence permutability we obtain the further consequences:

3. By 1.4-Theorem 6 every non-trivial algebra of a congruence distributive variety generated by a two-element algebra is isomorphic to a subdirect power of the generating two-element algebra, i.e. there is exactly one subdirectly irreducible algebra in the variety. As shown in 1.5, there are at most two subdirectly irreducible algebras in a variety generated by a two-element paraprimal algebra. In $V_{\underline{O_4}}$ there exist two subdirectly irreducible algebras. Adding nullary functions to the fundamental operations one get a subdirectly irreducible algebra which is not isomorphic to the generating algebra of the variety. The cardinality of this subdirectly irreducible algebra agrees with the cardinality of the generating algebra of the variety. The other varieties contain at most three non-isomorphic subdirectly irreducible algebras. These algebras have at most three elements. Therefore we have proved the following proposition: If $\underline{A}$ is any two-element algebra, the variety $V_{\underline{A}}$ contains at most three subdirectly irreducible algebras and any subdirectly irreducible algebra in $V_{\underline{A}}$ has at most three elements (for this result see also Taylor [3] and Berman [1]).

4. The remarks in 3. give rise to the following considerations about the finite algebra in a variety, especially the fine spectrum (Taylor 2), i.e. the number $f_V(n)$ of non-isomorphic algebras of the cardinality n in V.

Every finite Boolean algebra is a direct power of the two-element Boolean algebra $\underline{2_B}$. For the variety $V_{\underline{2_B}}$ of Boolean algebras it holds

$$f_{V_{\underline{2_B}}}(n) = \begin{cases} 1 \text{ if } n \text{ is a power of } 2 \\ 0 \text{ otherwise.} \end{cases}$$

Let $\underline{A}$ be a finite (k-element) non-trivial algebra. Under which condition does the variety generated by $\underline{A}$ have the fine spectrum

$$f_{V_{\underline{A}}}(n) = \begin{cases} 1 \text{ if } n \text{ is a power of } k \\ 0 \text{ otherwise ?} \end{cases}$$

Since every finite algebra in a variety generated by a primal algebra is a direct power of the generating algebra, the variety generated by a primal algebra has this property. To formulate Taylor's results we give two definitions. Let $\underline{A}$ be a two-element paraprimal algebra. If $\underline{A}$ is quasiprimal or $\underline{A}$ has a trivial subalgebra, then for a finite algebra $\underline{B}$ it holds $\underline{B} \in V_{\underline{A}}$ if and only if $\underline{B} \in IP\,\underline{A}$, i.e. $f_{V_{\underline{A}}}(n) = 0$ if n is a power of $\underline{A}$. We say that such paraprimal algebras are of first kind. The two-element paraprimal algebras $\underline{C}_1, \underline{C}_3, \underline{C}_4, \underline{D}_3, \underline{D}_1$ are of first kind since these algebras are quasiprimal; $\underline{L}_3, \underline{L}_4$ are also paraprimal algebras of first kind since $\underline{L}_3$ and $\underline{L}_4$ have one-element subalgebras. $\underline{L}_5$ and $\underline{L}_1$ are paraprimal, but they are not quasiprimal and they have no proper subalgebras. We say that such paraprimal algebras are of second kind. For paraprimal algebras of second kind it holds 1.5-Theorem 9. Now the result of Taylor ([21]) can be formulated in the following manner:

<u>Theorem 2:</u> Let V be a variety which is generated by its finite elements, then V has the fine spectrum

$$f_V(n) = \begin{cases} 1 \text{ if } n \text{ is a power of } 2 \\ 0 \text{ otherwise} \end{cases}$$

if and only if V is generated by a two-element paraprimal algebra of first kind and V has the fine spectrum

$$f_V(n) = \begin{cases} 1 \text{ if } n=1 \\ 2 \text{ if } n \text{ is a power of } 2 \\ 0 \text{ otherwise} \end{cases}$$

if and only if V is generated by a two-element paraprimal algebra of second kind.

The knowledge of the fine spectrum f_V refines that of the spectrum of V. The spectrum of a variety V, (or of an arbitrary class of algebras), is the set of the cardinalities of the finite members of V:

spec V = $\{ n \in N: f_V(n) > 0 \}$,(N is the set of the natural numbers).
Let $\underline{A} = \langle A,F \rangle$,(card A = k, k > 1), be a finite non-trivial algebra. $V_{\underline{A}}$ is said to have a minimal spectrum if spec $V_{\underline{A}} = \{ k^n \mid n \geq 0 \}$.
If $V_{\underline{A}}$ has a minimal spectrum, then $f_{V_{\underline{A}}}(n) = 0$ if n is not a power of k. In ([3]) Quackenbush has characterized all varieties with a minimal spectrum. $V_{\underline{A}}$ with $\underline{A} = \langle A,F \rangle$,(A = $\{0,\ldots,k-1\}$, k > 1) has a minimal spectrum if and only if $\underline{A}$ is paraprimal and all proper subalgebras of $\underline{A}$ are trivial.

If $V_{\underline{A}}$ has a minimal spectrum, then either
$f_{V_{\underline{A}}}(k^m) = 1$, (m > 0),
(in particular if $\underline{A}$ is quasiprimal or has a trivial subalgebra) or
$f_{V_{\underline{A}}}(k^m) = 2$,
(the diagonal subalgebra of $\underline{A}^2$ is a congruence class of a congruence on $\underline{A}^2$). If k is not a prime power , then $\underline{A}$ is quasiprimal. If $\underline{A}$ is not quasiprimal, then the diagonal subalgebra of $\underline{A}^2$ is a congruence class of a congruence on $\underline{A}^2$.

Concerning varieties generated by two-element algebras we have exactly 9 varieties with minimal spectrum: $V_{\underline{C_1}}$, $V_{\underline{C_3}}$, $V_{\underline{C_4}}$, $V_{\underline{D_3}}$, $V_{\underline{D_1}}$, $V_{\underline{L_1}}$, $V_{\underline{L_4}}$, $V_{\underline{L_5}}$.

In 4.3 we shall investigate independent sets of preprimal algebras. Here we give the definition of an independent set and show that in some two-element algebras the carrier is independent.

5. In ([1]) E.Marczewski defined the notion of an independent set in a universal algebra as a generalization of the notion of independence in various mathematical branches.

Definition 9: Let $\underline{A} = \langle A,F \rangle$ be an algebra and $I \subseteq A$, $I \neq \emptyset$, a subset of A. Then I is called an independent set of $\underline{A}$ if every function f of[I] into A can be extended to a homomorphism of $\underline{[I]}$ into $\underline{A}$ where $\underline{[I]}$ is the subalgebra of $\underline{A}$ generated by I.

The terminology „independent set" is justified by the following property of such sets: For any pairwise distinct elements $a_1,\ldots,a_n \in I$ and any term functions f,g of $\underline{A}$ it holds:
If $f(a_1,\ldots,a_n) = g(a_1,\ldots,a_n)$, then $f(x_1,\ldots,x_n) = g(x_1,\ldots,x_n)$ for all $x_1,\ldots,x_n \in A$.

Elements of the centroid $C(\underline{A})$ of $\underline{A}$ (2.2) cannot be independent; in fact we have:

<u>Lemma 3 (Marczewski [1]):</u> If I is independent in $\underline{A}$, then $I \wedge C(\underline{A}) = \emptyset$.

Let us consider algebras with independent carriers.

<u>Lemma 4 (Swierczkowski [1]):</u> Each algebra $\underline{A} = \langle A,F \rangle$, card $A \geq 3$, the carrier of which is independent in $\underline{A}$, has the form $\langle A,\emptyset \rangle$, i.e. the set of term functions is the empty set.

Obviously, for trivial algebras the carrier is independent. Which non-trivial algebras have independent carriers ?
By Lemma 4 it remains only the case of two-element algebras. From the set of all two-element algebras we pick out those with independent carriers. Such an algebra cannot have a constant term function, but it has two one-element subalgebras. All term functions of this algebra are self-dual, i.e.

$Nf(x_1,\ldots,x_n) = f(Nx_1,\ldots,Nx_n)$.

These conditions are only satisfied by $\underline{D_1} = \langle \{0,1\}, x+y+z, xy \vee xz \vee yz \rangle$, $\underline{L_4} = \langle \{0,1\}, x+y+z \rangle$, and $\underline{D_2} = \langle \{0,1\}, xy \vee xz \vee yz \rangle$. Therefore the algebras $\underline{D_1}$, $\underline{D_2}$, $\underline{L_4}$ have independent carriers.

3.3 Completeness properties of two-element algebras

In (Post[1],[3]) it is proved that a set of Boolean functions $F \subseteq \underline{F}(\{0,1\})$ is complete if and only if F is not included in one of the five maximal subclasses of $\underline{F}(\{0,1\})$. Thus the characterization of preprimal two-element algebras gives also conditions for a two-element algebra to be primal.

Recall that exactly the preprimal two-element algebras $\underline{C}_3(\underline{C}_2)$ and $\underline{D}_3$ generate arithmetical varieties, that $\underline{C}_3(\underline{C}_2)$ has a one-element subalgebra and that $\underline{D}_3$ has a non-trivial automorphism. Therefore a two-element algebra is primal if and only if it generates an arithmetical variety, has no proper subalgebras and no non-identical automorphism.

Recall that exactly the two-element algebras $\underline{C}_3(\underline{C}_2)$, $\underline{D}_3$, $\underline{L}_1$ generate congruence permutable varieties and that $\underline{L}_1$ is linear. Therefore a two-element algebra is primal if and only if it generates a congruence permutable variety, has no proper subalgebras, no non-identical automorphism and is not linear.

Only the preprimal algebra $\underline{L}_1$ does not generate a congruence distributive variety. Thus a two-element algebra is primal if and only if it generates a congruence distributive variety, has no proper subalgebras, no non-identical automorphism and is not monotonic.

Analogously we obtain conditions for a functionally complete two-element algebra. The algebra $\underline{A} = \langle\{0,1\},F\rangle$ is functionally complete if and only if the algebra $\underline{A}^+ = \langle\{0,1\},F^+\rangle$ is primal. Here $\underline{A}^+$ is the algebra obtained from $\underline{A}$ by making every element from $\{0,1\}$ the value of a new nullary term function.

Since $F^+ \nsubseteq C_3(C_2)$, $F^+ \nsubseteq D_3$, a two-element algebra is functionally complete if and only if it is not linear and not monotonic.

A two-element algebra which generates a congruence permutable variety is functionally complete if and only if it is not linear and a two-element algebra which generates a congruence distributive variety is functionally complete if and only if it is not monotonic.

Problems and comments

1. Give a new proof of Post's proposition that $\underline{A_1}$, $\underline{C_3}$, $\underline{C_2}$, $\underline{L_1}$, $\underline{D_3}$ are exactly all two-element preprimal algebras (up to equivalence) applying the ideas of Quackenbush's new proof of Rosenberg's theorem (see Appendix).

2. (Berman [2]) Provide a direct proof that there exist only countably many inequivalent varieties generated by a two-element set. Here „direct" means avoiding an enumeration of all classes as done by Post.

3. Which two-element algebras have non-trivial automorphisms and weak automorphisms (see 4.3) ?

4. Preprimal algebras

Now we consider preprimal algebras with more than two elements. By Definition 6 for a preprimal algebra $\underline{A} = \langle A,F \rangle$ we have the following alternative: Either not all nullary functions are term functions of $\underline{A}$ or $\underline{A}$ is constantive; in the first case $\underline{A}$ is functionally complete since F is a maximal class in $\underline{F}(A)$. Constantive algebras have no proper subalgebras and no non-identical automorphisms. Functionally complete preprimal algebras are simple. Recall that $\underline{C_3}$ and $\underline{D_3}$ are functionally complete two-element preprimal algebras, $\underline{A_1}$ and $\underline{L_1}$ are constantive. $\underline{C_3}$ and $\underline{D_3}$ are quasiprimal and it arises the question: Is every functionally complete preprimal algebra quasiprimal ?

The classification of all maximal classes in $\underline{F}(A)$ (card $A > 2$) by Rosenberg is the starting point of our next considerations. From this classification we obtain all forms of preprimal algebras.

4.1 Classification of all maximal subclasses of $\underline{F}(A)$

In ([3]) Post has given all maximal subclasses of $\underline{F}(A)$ if card $A = 2$. Jablonskij ([1]) has described all maximal subclasses in $\underline{F}(A)$ if card $A = 3$ and some maximal subclasses if card $A > 3$. In (Rosenberg[2]) an important but very difficult theorem is proved. This theorem describes all maximal classes of functions on a finite set with more than two elements in terms of the relational systems O,P,L,C,E,B.

O is the system of all order relations ϱ on A, i.e. reflexive, antisymmetric, and transitive binary relations, with a least and a greatest element.

P is the system of all binary relations $\{(a,s(a)) \mid a \in A\}$ where s is a permutation of A without invariant elements with all cycles of the same prime length.

L is the system of all quaternary relations λ defined by $(a,b,c,d) \in \lambda$ if and only if $a+b = c+d$, here $+$ is the addition of an elementary abelian p-group on A.

C is the system of all central, totally reflexive, and totally symmetric h-ary relations τ with $1 \leq h \leq \text{card } A$. Totally reflexive relations were defined in 1.1. An h-ary relation τ is said to be totally symmetric if for any permutation s of $\{0,\dots,h-1\}$ $(x_0,\dots,x_{h-1}) \in \tau$ implies $(x_{s(0)},\dots,x_{s(h-1)}) \in \tau$ $(1 \leq h \leq \text{card } A)$. The centre of a totally symmetric and reflexive h-ary relation τ is the set of all elements $c \in A$ such that $(c,x_1,\dots,x_{h-1}) \in \tau$ for all $x_1,\dots,x_{h-1} \in A$. A relation is said to be central if it is totally reflexive and symmetric and has a non-empty centre $\neq A$.

B is the system of all relations described in the following steps according to Rosenberg ([2]) and Knoebel ([5]).
Assume $h \geq 3$. A primitive h-elementary relation Υ is the h-ary relation on the set $E_h = \{0,\dots,h-1\}$ such that $(a_0,\dots,a_{h-1}) \in \Upsilon$ if at least two of the a_i are equal. Assume $m \geq 1$. An h-elementary relation η is the h-ary relation on the set E_h^m of all m-tuples over E_h such that
$((a_{00},\dots,a_{0m-1}),\dots,(a_{h-10},\dots,a_{h-1m-1})) \in \eta$
if for each $j < m$, at least two of the elements $a_{0j},a_{1j},\dots,a_{h-1j}$ are equal. A relation η is the m-fold power of the primitive h-elementary relation Υ. Let A be any finite set with at least h^m elements and let $\Phi : A \longrightarrow E_h^m$ be a function from A onto E_h^m. We define an h-ary relation δ on A by
$(a_0,\dots,a_{h-1}) \in \delta$ iff $(\Phi(a_0),\dots,\Phi(a_{h-1})) \in \eta$.
We will say that δ is an h-universal relation.

E is the system of all non-trivial equivalence relations Θ $(\Theta \neq O, \Theta \neq I)$.

Then we have the following proposition: A preprimal algebra is an algebra generated by an arbitrary relation from one of the relational classes O,P,L,C,B,E. Consequently, there are exactly the following classes of preprimal algebras:

1. Algebras $\underline{A}_\varsigma = \langle A,F_\varsigma\rangle$ generated by a relation ς from O,
2. Algebras $\underline{A}_s = \langle A,F_s\rangle$ generated by a relation $\{(a,s(a))|a \in A\}$ from P,
3. Algebras $\underline{A}_\alpha = \langle A,F_\alpha\rangle$ generated by a relation α from L,
4. Algebras $\underline{A}_\tau = \langle A,F_\tau\rangle$ generated by a relation τ from C with $2 \leq h \leq \mathrm{card}A$,
5. Algebras $\underline{A}_B = \langle A,F_B\rangle$ generated by a relation B from C with h = 1.
   The term functions of an algebra $\underline{A}_B$ are exactly all functions over A which preserve a subset $\emptyset \neq B \subset A$ of A.
6. Algebras $\underline{A}_\theta = \langle A,F_\theta\rangle$ generated by a relation θ from E,
7. Algebras $\underline{A}_\delta = \langle A,F_\delta\rangle$ generated by a relation δ from B.

A certain difficulty concerns the declaration of the type, i.e. the specification of the fundamental operations having prescribed numbers of arguments. In many investigations of preprimal algebras it is sufficient to think of an algebra as one in the Marczewski sense, i.e. as a collection of operations on a set. But for the equational aspect of preprimal algebras it is essential how the algebras are presented. Since a preprimal algebra is defined by a relation preserving the term functions, in many cases we do not know simple or natural generating subsets of the term functions. Therefore as an index set for our fundamental operations we just take the set Ω of all term functions of the algebra $\underline{A}$. We also take Ω to be the set of terms which will form the words. For any algebra $\underline{B}$ in $V_{\underline{A}}$ and for any term $\underline{f} \in \Omega$ we write $f^{\underline{B}}$ as the interpretation of $\underline{f}$ in $\underline{B}$. $\underline{B}$ can be written as $\langle B,\Omega^{\underline{B}}\rangle$, where $\Omega^{\underline{B}}$ is the indexed family $\{f^{\underline{B}}\}_{f\in\Omega}$. For $\underline{A}$ this means that the operations are indexed by themselves and thus $\Omega^{\underline{A}}$ is the identity on Ω. Note that this specification gives an infinite type. D.Lau ([2]) has shown that the preprimal algebras $\underline{A}_s$, $\underline{A}_\alpha$, $\underline{A}_\tau$, $\underline{A}_B$, $\underline{A}_\theta$, $\underline{A}_\delta$ are finitely presentable.

Let us add some historical remarks on the concept of a preprimal algebra. The beginning of the young history of the algebras $\underline{A}_\varsigma$ was their discovery by Jablonskij ([1]). Then Martynjuk ([1]) showed that they are preprimal. The equational maximality for all algebras generated by an arbitrary relation $\varrho \in O$ was found by Knoebel ([2]) and independently by the present author (Denecke[1]). The algebras $\underline{A}_s$ were discovered by Jablonskij ([1]). Quackenbush ([1]) introduced the class of demiprimal algebras which includes all algebras $\underline{A}_s$. Jablonskij ([1]) and Rosenberg ([1]) defined the algebras $\underline{A}_\alpha$. In the more general framework of paraprimal algebras Clark and Krauss ([2]) established a structure theorem for the variety $V_{\underline{A}_\alpha}$ generated by $\underline{A}_\alpha$. Preprimal algebras $\underline{A}_B$ were found by Jablonskij ([1]). $\underline{A}_B$ is a semiprimal algebra of Foster/Pixley ([1],[2]). Rosenberg ([1]) defined the algebras $\underline{A}_\tau$ $(h > 1)$. Knoebel ([3]) and the author ([1]) independently discovered their equational maximality as well as the fact that $\underline{A}_\tau$ $(h > 1)$ is the only subdirect irreducible one in the variety generated. The algebras $\underline{A}_\theta$ were defined by Jablonskij ([1]). $\underline{A}_\theta$ is one of the linear hemiprimal algebras of Foster ([8]) and therefore all algebras $\underline{A}_{\hat{\theta}}$ are included in the larger class of arithmetical algebras (Pixley[5]). The algebras generated by the so-called primitive h-elementary relations were first defined by Jablonskij ([1]). They also figure prominently in Butler's ([1]) criterion for functional completeness. Martynjuk ([1]) generalized them and Rosenberg ([1]) further enlarged this class to complete his sixth class. Knoebel ([5]) discovered that there are only finitely many subdirect irreducibles in $V_{\underline{A}_\delta}$ and characterized the subvarieties.

4.2 Elementary properties of preprimal algebras

A preprimal algebra is either functionally complete or constantive. We now pay attention to the question: Is every functionally complete preprimal algebra also quasiprimal ? In 3.1 we have remarked that every two-element preprimal algebra is either quasiprimal or constantive. As a generalization of this result we obtain:

Theorem 1: A preprimal algebra is either quasiprimal or constantive.

Proof: Let ϱ be an h-ary relation ($h \geq 2$) with $(a,\dots,a) \in \varrho$ for all $a \in A$, i.e. the diagonal $\Delta(\underline{A})$ is a subset of ϱ. If c_i is any constant function of $\underline{F}(A)$, ($c_i(x) \equiv i, i \in A$ for all $x \in A$) then $(a_1,\dots,a_h) \in \varrho$ implies $(c_i(a_1),\dots,c_i(a_h)) = (i,\dots,i) \in \varrho$. Therefore all constant functions of $\underline{F}(A)$ preserve the relation ϱ. For instance, for all totally reflexive h-ary relations with $2 \leq h \leq \text{card } A$ it holds $(a,\dots,a) \in \varrho$, i.e. all constant functions of $\underline{F}(A)$ preserve a totally reflexive h-ary relation ($2 \leq h \leq \text{card } A$). All relations from O are reflexive. The relations from L contain all (a,a,a,a). The relations from C are totally reflexive if $h \geq 2$. The relations from B are totally reflexive. Since the permutation s has no invariant elements, the relations from P are not reflexive. The algebras $\underline{A_B}$ (relations from C with h=1) contain not all constant functions as term functions since a constant function with a value not contained in B does not preserve the set B. We have proved that the preprimal algebras of the forms $\underline{A_\varrho}$, $\underline{A_\alpha}$, $\underline{A_\tau}$, $\underline{A_\theta}$, $\underline{A_\delta}$ are constantive while preprimal algebras of the forms $\underline{A_B}$, $\underline{A_s}$ are not constantive. Preprimal algebras $\underline{A_B}$, $\underline{A_s}$ are functionally complete. By Definition 1 and by Definition 2 algebras $\underline{A_B}$ are semiprimal (more precisely subprimal) and algebras $\underline{A_s}$ are demiprimal. Semi- and demiprimal algebras are quasiprimal.

By 2.1-Corollary 2 and Theorem 1 it follows:

Corollary 2: Every preprimal algebra is functionally complete if and only if it is quasiprimal.

By Lausch/Nöbauer ([1]) for any algebra there are the following cases:

1. $\underline{A}$ is functionally complete,

2. $\underline{A}$ is functionally semicomplete, i.e. every unary function is a polynomial function of $\underline{A}$, but there is no natural number $n > 1$ such that every n-ary function is a polynomial function on $\underline{A}$.

3. $\underline{A}$ is functionally incomplete, i.e. there is no natural number n, $n>0$, such that every n-ary function is a polynomial function on $\underline{A}$.

Applying this to preprimal algebras we obtain the following classes of preprimal algebras:

functionally complete preprimal algebras: $\underline{A_B}$, $\underline{A_s}$

functionally incomplete preprimal algebras: $\underline{A_\xi}$, $\underline{A_\alpha}$, $\underline{A_\tau}(h > 1)$, $\underline{A_\theta}$, $\underline{A_\delta}$ (δ is a non-primitive h-elementary relation),

functionally semicomplete preprimal algebras: $\underline{A_\perp}$.

Any constantive algebra has no proper subalgebras and no non-identical automorphisms. A demiprimal algebra has no proper subalgebras (2.2-Theorem 6) and a semiprimal algebra has no non-identical automorphisms (2.2-Theorem 3). Therefore we obtain:

<u>Corollary 3:</u> A preprimal algebra which is not of the form $\underline{A_B}$ has no proper subalgebras and a preprimal algebra which is not of the form $\underline{A_s}$ has no non-identical automorphisms.

Which preprimal algebras are simple ? Obviously, $\underline{A_B}$ and $\underline{A_s}$ are simple since these algebras are quasiprimal. By Definition 4 algebras $\underline{A_\theta}$ are hemiprimal, more precisely, linear hemiprimal. The algebras $\underline{A_\theta}$ have only the non-trivial congruence relation θ. We obtain:

<u>Corollary 4:</u> Every preprimal algebra which is not of the form $\underline{A_\theta}$, is simple.

<u>Proof:</u> If a preprimal algebra $\underline{A} = \langle A,F\rangle$, distinct from A_θ, would have a non-trivial congruence relation θ, all term functions of $\underline{A}$ preserve θ, i.e. $F \subseteq F_\theta$. This contradicts our assumption that F is a maximal class in $\underline{F}(A)$. ▲

Algebras of the form $\underline{A_B}$ have, as subprimal algebras, exactly one subalgebra which is trivial or primal. We investigate the non-trivial homomorphic image of an algebra $\underline{A_\theta}$. The algebras $\underline{A_\theta}$ are constantive. Is the non-trivial homomorphic image of $\underline{A_\theta}$ also constantive ? To prove this conjecture we apply a proposition by Foster.

Lemma 5 (Foster [7]): Let $\underline{A}$, $\underline{B}$ be algebras of the same type and with the identities of $\underline{A}$ contained in those of $\underline{B}$: $\mathrm{Id}\ \underline{A} \subseteq \mathrm{Id}\ \underline{B}$. Then $C(\underline{B})$, the centroid of $\underline{B}$, is a homomorphic image of $C(\underline{A})$, the centroid of $\underline{A}$: $C(\underline{A}) \twoheadrightarrow C(\underline{B})$. In particular, if $\mathrm{Id}\ \underline{A} = \mathrm{Id}\ \underline{B}$, then $C(\underline{A})$ is isomorphic to $C(\underline{B})$.

Lemma 6: The non-trivial homomorphic image of $\underline{A_\theta}$ is a primal algebra.
Proof: Since $\mathrm{Id}\ \underline{A_\theta/\Theta} \supseteq \mathrm{Id}\ \underline{A_\theta}$, by Lemma 5 it holds $C(\underline{A_\theta}) \twoheadrightarrow C(\underline{A_\theta/\Theta})$. $\underline{A_\theta}$ is constantive, therefore $C(\underline{A_\theta/\Theta})$ is a homomorphic image of $\underline{A_\theta}$. $C(\underline{A_\theta/\Theta})$ is a minimal subalgebra of $\underline{A_\theta/\Theta}$. The inverse homomorphic image of $C(\underline{A_\theta/\Theta})$ under the homomorphism $\underline{A_\theta} \twoheadrightarrow \underline{A_\theta/\Theta}$ is a subalgebra of $\underline{A_\theta}$. $\underline{A_\theta}$ has, as a constantive algebra, no proper subalgebras. Therefore $C(\underline{A_\theta/\Theta}) = \underline{A_\theta/\Theta}$ and $\underline{A_\theta/\Theta}$ is constantive, thus $\underline{A_\theta/\Theta}$ is also rigid. $\underline{A_\theta/\Theta}$ is simple and generates an arithmetical variety since $\underline{A_\theta}$ is linear hemiprimal and $\mathrm{Id}\ \underline{A_\theta/\Theta} \supseteq \mathrm{Id}\ \underline{A_\theta}$. By 2.2-Corollary 7 $\underline{A_\theta/\Theta}$ is primal. ▲

An algebra $\underline{A_s}$ has the automorphism determined by the permutation s without invariant elements with cycles of the same prime length. The other automorphisms of $\underline{A_s}$ are determined by the powers of s. The automorphism group of $\underline{A_s}$ is cyclic and of prime order. We obtain:

Theorem 7: Every proper non-trivial homomorphic image and every proper non-trivial subalgebra of a preprimal algebra is primal. The non-trivial automorphism group of a preprimal algebra is cyclic and of prime order.

We give a survey on the results of 4.2:

quasiprimal: $\underline{A_s}$, $\underline{A_B}$

rigid: $\underline{A_\varrho}$, $\underline{A_B}$, $\underline{A_\theta}$, $\underline{A_\alpha}$, $\underline{A_\tau}$, $\underline{A_\delta}$

constantive: $\underline{A_\varrho}$, $\underline{A_\theta}$, $\underline{A_\alpha}$, $\underline{A_\tau}$, $\underline{A_\delta}$

simple: $\underline{A_\varrho}$, $\underline{A_s}$, $\underline{A_B}$, $\underline{A_\alpha}$, $\underline{A_\tau}$, $\underline{A_\delta}$

minimal: $\underline{A_\varrho}$, $\underline{A_s}$, $\underline{A_\theta}$, $\underline{A_\alpha}$, $\underline{A_\tau}$, $\underline{A_\delta}$

4.3 Independent sets and free preprimal algebras

Now we will consider independent sets in preprimal algebras and we will characterize free preprimal algebras. Since a primal algebra $\underline{A} = \langle A,F\rangle$ is constantive, $\underline{A}$ has, according to 3.2-Lemma 3, no independent sets. Is there a preprimal algebra $\underline{A} = \langle A,F\rangle$ with independent sets ? By 3.2-Lemma 3 and 4.2-Theorem 1 we can restrict our investigations to quasiprimal preprimal algebras. An independent set of $\underline{A_B}$ does not contain any element of B, i.e. $I \subseteq A\setminus B$ (see also 2.2). Let us consider the following cases:

1. B contains at least two distinct elements.

Let f be a function $f: I \longrightarrow A, a \in I, b \in B$ with $f(a) = b$, let φ be a unary term function of $\underline{A_B}$ with $\varphi(a) = a$, $\varphi(b) = c \in B$, $b \neq c$, $\varphi(x) \in B$ if $x \neq a$. Since φ preserves B, φ is a term function of $\underline{A_B}$. Then it follows $\varphi(f(a)) = \varphi(b) = c$, but $f(\varphi(a)) = f(a) = b \neq c$. Therefore it is impossible to extend the function f to a homomorphism of $\underline{[I]}$ into $\underline{A_B}$, thus I is not independent.

2. $B = \{a\}$, I contains at least two distinct elements.

If $I \subseteq A\setminus\{a\}$ is an independent set in $\underline{A_B}$ and $b,c \in I, b \neq c$, f is a function $f: I \longrightarrow A$ with $f(c) = a \in B$, $f(b) = c \in I$, $(a \neq c)$ and let φ be a unary term function of $\underline{A_B}$ with $\varphi(c) = b$, $\varphi(a) = a$, then $f(\varphi(c)) = f(b) = c$, $\varphi(f(c)) = \varphi(a) = a \neq c$. Therefore f cannot be extended to a homomorphism of $\underline{[I]}$ into $\underline{A_B}$.

3. $B = \{a\}$, $I = \{b\}$, $k \geq 4$.

Let f be a function $f: I \longrightarrow A$ with $f(b) = c \neq a$, let φ be a unary term function of $\underline{A_B}$ with $\varphi(a) = a$, $\varphi(b) = b$, $\varphi(c) = d \neq c$, then $\varphi(f(b)) = d$, but $f(\varphi(b)) = f(b) = c$.

4. $B = \{a\}$, $I = \{b\}$, $k = 3$.

Let $A = \{a,b,c\}, (a \neq b \neq c), f: I \longrightarrow A$ with $f(b) = c$. Since the function g with $g(a) = a$, $g(b) = c$, $g(c) = b$ is a term function of $\underline{A}$ and c_a with $c_a(x) \equiv a$ is also a term function of $\underline{A}$, it holds $\underline{[\{b\}]} = \underline{A}$. Only the one-element algebra with the carrier $\{a\}$ or A are possible as homomorphic images of $\underline{[\{b\}]} = \underline{A}$. If it is possible to extend f to a homomorphism h into $\underline{A}$, it follows $h(a) = a$ and $h(c) = b$ or $h(a) = b$ and $h(c) = a$. The function φ with $\varphi(a) = a$, $\varphi(b) = b$, $\varphi(c) = b$ is a term function of $\underline{A}$. It holds $h(\varphi(c)) = h(b) = c$, but $\varphi(h(c)) = \varphi(b) = b \neq c$. In the latter case we obtain $h(\varphi(c)) = h(b) = c$, but $\varphi(h(c)) = \varphi(a) = a \neq c$.

5. $B = \{a\}$, $I = \{b\}$, $k = 2$.

We investigate the preprimal algebra $\underline{C_3} = \langle \{0,1\}, +, \wedge, 0\rangle$. Only $I = \{1\}$ is possible as independent set. We consider a function $f: 1 \longrightarrow 0$ and we realize that f can be extended to a homomorphism of $\underline{[\{1\}]} = \underline{C_3}$ into $\underline{C_3}$. The homomorphic image of $\underline{[\{1\}]}$ is the one-element algebra $\langle \{0\}, +, \wedge, 0\rangle$. Now we consider a function $f: 1 \longrightarrow 1$ and we realize that f can be extended to a homomorphism, namely to the identical automorphism. Therefore $\{1\}$ is an independent set in $\underline{C_3}$. We obtain:

<u>Lemma 8:</u> There is (up to equivalence) exactly one preprimal algebra of the form $\underline{A_B}$ which has an independent set: $\underline{C_3} = \langle \{0,1\}, +, \wedge, 0\rangle$.

<u>Lemma 9:</u> Preprimal algebras of the form $\underline{A_s}$ have independent sets if and only if card A is a prime number. These independent sets are exactly all one-element subsets of A.

<u>Proof:</u> At first we have to prove that an algebra $\underline{A_s}$ has no independent sets with more than one element. Let $I \subset A$ be an independent set with a, $b \in I$, $a \neq b$, $c \notin I$ (there exists such an element c since algebras in which the carriers are independent have not the form $\underline{A_s}$). $\underline{A_s}$ has, as a demiprimal algebra, no proper subalgebras, consequently $\underline{[I]} = \underline{A_s}$. $\underline{A_s}$ is simple, therefore a homomorphic image is either a one-element algebra or $\underline{A_s}$. In the second case the homomorphism is an automorphism. We consider a function f with $f(a) = a$, $f(b) = c$. Since the automorphisms of $\underline{A_s}$ have no fixed elements or they are the identical automorphism, it is impossible to extend this function to an automorphism of $\underline{A_s}$. Since $c \neq a$ it is also impossible to extend f to a homomorphism of a one-element algebra. Let $I = \{a\}$ be an arbitrary one-element subset of A and let card A be a prime number. We extend a function f with $f(a) = a$ to the identical automorphism. For any $b \in A \setminus I$ there exists a power of s with

$s^l(a) = b$ since s is representable by exactly one cycle of prime length. Therefore a function f with $f(a) = b$ can be extended to an automorphism of $\underline{A_s}$. If card A is not a prime number, s is representable by a product of cycles with the same prime length. In this case card $A > 3$. Let a be any element of A. If a and b are in distinct cycles of s, then for every power s^l of s it holds $s^l(a) \neq b$. Therefore a function f with $f(a) = b$ cannot be extended to an automorphism of $\underline{A_s}$. $\underline{A_s}$ does not contain an independent set. ▲

We proved that independent sets exist only in the preprimal algebras $\underline{A_s}$ (card A prime number) and in $\underline{C_3} = \langle \{0,1\}, +, \wedge, 0\rangle$. These are one-element independent sets. In $\underline{A_s}$ (card A prime) any one-element subset of A is independent, in $\underline{C_3} = \langle \{0,1\}, +, \wedge, 0\rangle$ only the set $\{1\}$ is independent. These one-element independent sets generate the algebras $\underline{A_s}$ and $\underline{C_3}$, i.e. they are basis sets of the algebras. If $I \subseteq A$ is independent in $\underline{A}$, we can extend any function $f: I \dashrightarrow A$ to a homomorphism in any algebra from $V_{\underline{A}} = HSP\underline{A}$. I is said to be independent with respect to $V_{\underline{A}}$. An algebra $\underline{A}$ is free in $V_{\underline{A}}$ if $\underline{A}$ is generated by a set I, $\emptyset \neq I \subseteq A$, which is independent with respect to $V_{\underline{A}}$. $\underline{A}$ is called a free algebra if $\underline{A}$ is free in the class $\{\underline{A}\}$. We obtain:

<u>Theorem 10:</u> The algebras $\underline{A_s}$ (card A is a prime number) and $\underline{C_3}$, $\underline{C_3} = \langle \{0,1\}, +, \wedge, 0\rangle$ are (up to equivalence) all free preprimal algebras.

4.4 Weak automorphisms of preprimal algebras

In the sections 4.2 and 4.3 preprimal algebras were investigated up to equivalence and the types of the algebras were neglected. The notion of a weak automorphism of a universal algebra is also considered in the category of non-indexed algebras. This is a motivation to apply the concept of a weak automorphism defined by A.Goetz ([1]) and E. Marczewski ([1]) and to investigate weak automorphisms of preprimal algebras. Now we consider another algebraic approach to maximal classes of k-valued functions, the maximal function algebras. In (A.I.Malcev[2]) the notion of a function algebra is defined. $\langle F, e_1^2, \zeta, \tau, \Delta, *\rangle$ is the function algebra corresponding to the algebra $\underline{A} = \langle A,F\rangle$. The carrier of the function algebra is the clone of $\underline{A} = \langle A,F\rangle$ and the operations are defined as follows:

$(\zeta f)(x_1,\dots,x_n) = f(x_2,x_3,\dots,x_n,x_1)$,

$(\tau f)(x_1,x_2,\dots,x_n) = f(x_2,x_1,\dots,x_n)$,

$(\Delta f)(x_1,\dots,x_{n-1}) = f(x_1,x_1,\dots,x_{n-1})$ if $n \geq 2$ and

$(\zeta f) = (\tau f) = (\Delta f) = f$ if $n = 1$.

$(g * f)(x_1,\dots,x_m,x_{m+1},\dots,x_{m+n-1}) = f(g(x_1,\dots,x_m),x_{m+1},\dots,x_{m+n-1})$,

$e_1^2(x_1,x_2) = x_1$ for all $x_1,\dots,x_{m+n-1} \in A$.

The algebra $\langle \underline{F}(A), e_1^2, \zeta, \tau, \Delta, *\rangle$ is called the full function algebra on A (iterative algebra in A.I.Malcev[2]). Each function algebra is a subalgebra of the full function algebra and the maximal subalgebras of the full function algebra correspond to preprimal algebras. In ([1], [2], [3]) D.Lau investigated properties of maximal function algebras.

Let Aut($\underline{A}$) be the automorphism group of $\underline{A}$. An automorphism of the function algebra corresponding to $\underline{A}$ is called a clone automorphism of $\underline{A}$. Let Aut(F) be the set of all clone automorphisms of $\underline{A}$. A permutation s of the carrier A of the algebra $\underline{A} = \langle A,F\rangle$ is said to be a weak automorphism if the deduced mapping

(1) $\alpha : f \longrightarrow f^{\alpha} = s \cdot f \cdot s^{-1}$

is a bijection of the set $\underline{F}$ (of all term functions of $\underline{A}$) onto $\underline{F}$, where the term function $s \cdot f \cdot s^{-1}$ is defined by

$$s \cdot f \cdot s^{-1}(x_1,\dots,x_n) = s(f(s^{-1}(x_1),\dots,s^{-1}(x_n)))$$

for all $x_1,\dots,x_n \in A$ and $n = 1,2,\dots$.

Of course, every automorphism of an algebra $\underline{A}$ is a weak automorphism of $\underline{A}$. Let $Aut^*(\underline{A})$ be the group of all weak automorphisms of $\underline{A}$. The mapping $\tilde{\alpha}: f \longrightarrow f^{\alpha} = s \cdot f \cdot s^{-1}$ is a clone automorphism of $\underline{A}$, i.e. an automorphism of the corresponding function algebra. Following A.I. Malcev ([3],[5]), the clone automorphisms defined by (1) are called inner automorphisms of the corresponding function algebra. Let $Q(\underline{A})$ be the set of all unary one-to-one term functions of $\underline{A}$. Since $\underline{A}$ is finite, $Q(\underline{A})$ is a group with superposition as group operation. It holds:

<u>Lemma 11 (Senft[1], Dudek/Glazek[1]):</u> Every unary one-to-one term function of a finite algebra $\underline{A} = \langle A,F \rangle$ is a weak automorphism of $\underline{A}$. $Q(\underline{A})$ is a normal subgroup of the group $Aut^*(\underline{A})$.

The mapping which assigns every weak automorphism s of $\underline{A}$ the corresponding inner clone automorphism $\alpha: s \longrightarrow \alpha$, is a group homomorphism of $Aut^*(\underline{A})$ into $Aut(\underline{F})$ with the kernel $Aut(\underline{A})$, i.e. the automorphisms of $\underline{A}$ are precisely those weak automorphisms with $s \longrightarrow 1_{\underline{F}}$, $1_{\underline{F}}$ is the unit-element of $Aut(\underline{F})$. Collecting these facts we have the following theorem:

<u>Theorem 12 (Senft[1]):</u> The mapping $s \longrightarrow \alpha$ is a homomorphism of $Aut^*(\underline{A})$ into $Aut(\underline{F})$ with the kernel $Aut(\underline{A})$. $Aut(\underline{A})$ is a normal subgroup of $Aut^*(\underline{A})$ and $Aut^*(\underline{A})/Aut(\underline{A})$ is isomorphic to a subgroup of $Aut(\underline{F})$.

By 4.2-Theorem 7 every non-trivial automorphism group of a preprimal algebra is cyclic and of prime order. Only preprimal algebras $\underline{A_s}$ have non-trivial automorphism groups. We are interested in the weak automorphisms of the algebras $\underline{A_s}$. Let $\underline{S_A}$ be the symmetric group on A, especially, let $\underline{S_k}$ be the symmetric group of the elements of $A=\{0,\dots,k-1\}$. By Theorem 12 $Aut(\underline{A_s})$ is a normal subgroup of $Aut^*(A_s)$ and by Lemma 11 $Q(\underline{A_s})$ is a normal subgroup of $Aut^*(\underline{A_s})$. $Q(\underline{A_s})$ contains precisely those

elements of $\underline{S_k}$ which are permutable with s, therefore $Q(\underline{A_s})$ is the normalizer of $\underline{[s]}$ in $\underline{S_k}$. Therefore $Q(\underline{A_s})$ is the normalizer of $Aut(\underline{A_s})$ in $\underline{S_k}$. The normalizer of $Aut(\underline{A_s})$ is the maximal subgroup of $\underline{S_k}$ in which $Aut(\underline{A_s})$ is a normal subgroup. Since $Aut(\underline{A_s}) \subseteq Q(\underline{A_s}) \subseteq Aut^*(\underline{A_s}) \subseteq \underline{S_k}$, we obtain $Q(\underline{A_s}) = Aut^*(\underline{A_s})$ or $Aut^*(\underline{A_s}) = \underline{S_k}$. In the second case $Aut(\underline{A_s})$ is a normal subgroup of $\underline{S_k}$. If $k \geqslant 5$, $\underline{S_k}$ has precisely the normal subgroups $\underline{E}, \underline{A_k}, \underline{S_k}$. ($\underline{E}$ is the unit-group, $\underline{A_k}$ is the group of all odd permutations). $\underline{A_k}$ is not cyclic for $k \geqslant 5$, therefore $Aut^*(\underline{A_s}) = \underline{S_k}$ if $k \geqslant 5$, thus $Aut^*(\underline{A_s}) = Q(\underline{A_s})$. If $k = 2$, then with $s = (01)$ we have:

$Aut(\underline{A_{(01)}}) = \underline{[(01)]} = Q(\underline{A_{(01)}}) = Aut^*(\underline{A_{(01)}}) = \underline{S_2}$.

If $k = 4$, then with $s_1 = (01)(23)$ or $s_2 = (02)(13)$ or $s_3 = (03)(12)$ $Aut(\underline{A_s})$ is isomorphic to the cyclic group of second order which is not a normal subgroup of $\underline{S_4}$, therefore $Aut(\underline{A_s}) \subset \underline{S_4}$ and $Q(\underline{A_s}) = Aut^*(\underline{A_s})$. If $k = 3$, then with $s = (012)$ we have $Aut(\underline{A_s}) = \underline{[(012)]} = \underline{A_3} = Q(\underline{A_s})$. In this case it holds $Aut^*(\underline{A_s}) = \underline{S_3}$. (see problems and complements)

We obtain:

Lemma 13: For every preprimal algebra of the form $\underline{A_s}$ $(k \neq 3)$ $Q(\underline{A_s}) = Aut^*(\underline{A_s})$, i.e. every weak automorphism is a unary one-to-one term function.

Corollary 14: The group of all weak automorphisms of $\underline{A_s}$ is the normalizer of s in $\underline{S_k}$.

Theorem 15: If $\underline{A}$ is any preprimal algebra, then $Aut(\underline{A})$ is a normal subgroup of $Q(\underline{A})$.

If $\underline{A}$ is a preprimal algebra of the form $\underline{A_s}$, Theorem 15 follows by Corollary 14. The other preprimal algebras have only identical automorphisms.

Is there a preprimal algebra $\underline{A}$ distinct from $\underline{A_s}$ with $Q(\underline{A}) = Aut^*(\underline{A})$? Let us consider algebras of the form $\underline{A_B}$. Let s be a permutation of A which does not preserve B, then there is an $a \in B$ and an element $c \notin B$ with $s(a) = c$. Let $s^{-1}(a) = b$, $b \neq a$, $b \in A$ and let c_a be a term function of $\underline{A_B}$ with $c_a(x) = a$. Then $s(c_a(s^{-1}(a))) = s(a) = c \notin B$. Therefore for a permutation s of A not preserving B $s \cdot f \cdot s^{-1}$ is no term function of $\underline{A_B}$, i.e. s is not a weak automorphism and $\mathrm{Aut}^*(\underline{A_B}) = Q(\underline{A_B})$.

Let s be a term function of A which is not a term function of $\underline{A_\Theta}$. Then there exist two elements $a \neq b \in A$ with $(a,b) \in \Theta$, but $(s(a),s(b)) \notin \Theta$. Let $s^{-1}(a) = d$, $s^{-1}(b) = r$, $r \neq d$, let f be the term function of $\underline{A_\Theta}$ defined by $f(x) = \begin{cases} a \text{ if } x=d \\ b \text{ otherwise} \end{cases}$. Then $s(f(s^{-1}(a))) = s(a)$, $s(f(s^{-1}(b)) = s(b)$ and $(s \cdot f \cdot s^{-1}(a), s \cdot f \cdot s^{-1}(b)) \notin \Theta$, i.e. $s \cdot f \cdot s^{-1}$ is no term function of $\underline{A_\Theta}$. Therefore $\mathrm{Aut}^*(\underline{A_\Theta}) = Q(\underline{A_\Theta})$.

On the other hand there exist preprimal algebras for which not every weak automorphism is a unary one-to-one term function. We consider an algebra of the form A_ϱ: $\underline{A} = \langle \{0,1,2\}, \max(x,y), \min(x,y), m_1, m_2, 0,1,2 \rangle$ with $m_1(x) = \begin{cases} 0 \text{ if } x=0 \\ 2 \text{ otherwise} \end{cases}$ and $m_2(x) = \begin{cases} 0 \text{ if } x=0 \text{ or } x=1 \\ 2 \text{ otherwise} \end{cases}$, (max, min with respect to the order on $\{0,1,2\}$ given by $0 \leq 1 \leq 2$). Since $\max(x,y)$, $\min(x,y), m_1(x), m_2(x), 0,1,2$ generate the class of all monotonic functions (Jablonskij [1]), $\underline{A}$ is preprimal. If $s = (02)$, $s \notin Q(\underline{A})$, then $s(m_1(s^{-1}(x))) = m_2(x)$, $s(m_2(s^{-1}(x))) = m_1(x)$, $s(\max(s^{-1}(x), s^{-1}(y))) = \min(x,y)$, $s(\min(s^{-1}(x), s^{-1}(y))) = \max(x,y)$. Therefore s is a weak automorphism of $\underline{A}$.

Theorem 12 describes a connection between the weak automorphisms of an algebra and the automorphisms of the corresponding function algebra, i.e. the clone automorphisms of $\underline{A}$. I.A.Malcev ([1]) proved that all automorphisms of a function algebra $\langle F, e_1^2, \zeta, \tau, \Delta, * \rangle$, $(F \subseteq \underline{F}(A))$, where F contains all constant functions of $\underline{F}(A)$, are inner automorphisms. D.Lau ([3]) proved that all automorphisms of the maximal classes in $\underline{F}(A)$ are inner automorphisms.
We obtain:

<u>Theorem 16:</u> If $\underline{A} = \langle A,F \rangle$ is an arbitrary preprimal algebra, then $\mathrm{Aut}(\underline{A})$ is isomorphic to $\mathrm{Aut}(\bar{F})$, i.e. the group of all weak automorphisms is isomorphic to the group of all clone automorphisms.

In (Lau[1]) congruences on maximal function algebras $\langle F, e_1^2, \zeta, \tau, \Delta, * \rangle$ are investigated. The result is that there are 3,4,6, or 7 congruences on these algebras.

For the investigation of preprimal algebras the results of Lau ([2]) on the order of maximal classes are of great interest. Let $F \subseteq \underline{F}(A)$, then the least number o such that F has a basis whose functions depend on at most o variables is called the order of this closed class. If o is finite, F has a finite basis and the algebra $\underline{A} = \langle A, F \rangle$ has a finite type. For the maximal classes of functions in $\underline{F}(A)$ it holds

$$o(F_s) = \begin{cases} 3 \text{ if card } A = 2 \\ 2 \text{ otherwise} \end{cases}, \quad o(F_\alpha) = o(F_\delta) = o(F_\theta) = 2, \quad 2 \leqslant o(F_\gamma) \leqslant \max(2,h),$$

$(1 \leqslant h \leqslant \text{card } A)$. For maximal classes F_ς it holds: If $2 \leqslant \text{card } A \leqslant 7$ or the supremum and the infimum, respectively, of two any elements of A with respect to ς are uniquely determined, then $o(F_\varsigma) = 2$. Therefore preprimal algebras $\underline{A_s}, \underline{A_\alpha}, \underline{A_\theta}, \underline{A_\delta}, \underline{A_B}, \underline{A_\gamma}$ have finite types. For $\underline{A_\varsigma}$, in generally, this problem is open.

<u>Corollary 17:</u> Preprimal algebras $\underline{A_s}, \underline{A_\alpha}, \underline{A_\theta}, \underline{A_\delta}, \underline{A_\gamma}, (2 \leqslant h \leqslant \text{card } A), \underline{A_B}$ have finite types.

Problems and comments

1. A finite algebra is said to be homogeneous if $\text{Aut}(\underline{A}) = \underline{S_A}$. B.Csakany ([2]) showed that, with the exception of six algebras, every homogeneous algebra is functionally complete. It is natural to consider weak homogeneous algebras, i.e. algebras with $\text{Aut}^*(\underline{A}) = \underline{S_A}$. Which properties have weak homogeneous algebras and which preprimal algebras are weak homogeneous ?
2. Prove that $\text{Aut}^*(\underline{A_s}) = \underline{S_3}$ if card A = 3.
3. Characterize the group of all weak automorphisms of a preprimal algebra.
4. For which preprimal algebras it holds $Q(\underline{A}) = \underline{S_A}$?
5. Does an algebra $\underline{A_\varsigma}$, (ς any relation from O) have a finite type ?

5. Varieties generated by preprimal algebras

5.1 An algebraic characterization of the preprimal algebras generating arithmetical varieties

Algebras of the forms $\underline{A_B}$, $\underline{A_s}$, $\underline{A_\theta}$ are semiprimal, demiprimal, and linear hemiprimal, respectively. They generate arithmetical varieties and there are no other preprimal algebras generating arithmetical varieties. By 4.2-Theorem 1 all preprimal algebras different from $\underline{A_B}$ and $\underline{A_s}$ are constantive, i.e. they have no proper subalgebras and they are rigid. All preprimal algebras different from $\underline{A_\theta}$ are simple. A simple algebra which has no proper subalgebras and no non-identical automorphisms and which generates an arithmetical variety is primal. From these remarks it follows:

<u>Proposition 1:</u> A preprimal algebra either generates an arithmetical variety or is constantive and simple.

A primal algebra $\underline{A}$ generates an arithmetical variety and it satisfies the conditions

a. $\underline{A}$ has no proper subalgebras,

b. $\underline{A}$ has no non-identical automorphisms, and

c. $\underline{A}$ is simple.

Algebras of the form $\underline{A_B}$ satisfy the conditions b.,c. while instead of a. it holds condition

a'. $\underline{A}$ has exactly one proper subalgebra which is primal or else trivial (2.2-Corollary 5).

Algebras of the form $\underline{A_s}$ satisfy the conditions a.,c. while instead of b. it holds condition

b'. $\underline{A}$ has a cyclic automorphism group of prime order.

Algebras of the form $\underline{A_\theta}$ satisfy the conditions a.,b. while instead of c. it holds condition

c'. $\underline{A}$ has exactly one non-trivial homomorphic image, this homomorphic image is a primal algebra.

A preprimal algebra which generates an arithmetical variety satisfies one of the three cases which we obtain substituting exactly one of the conditions a.,b.,c. by one of the conditions a'.,b'.,c'. and keeping fixed the remaining conditions. The converse is also true and therefore we have

<u>Theorem 2:</u> An algebra $\underline{A} = \langle A,F \rangle$,card $A > 1$, which generates an arithmetical variety is preprimal if and only if one of the following mutually excluding cases is satisfied:

1. a'. $\underline{A}$ has exactly one subalgebra $\underline{B}$ which is either primal or trivial,
   b. $\underline{A}$ is rigid,
   c. $\underline{A}$ is simple,
2. a. $\underline{A}$ has no proper subalgebras,
   b'. the automorphism group of $\underline{A}$ is cyclical and of prime order,
   c. $\underline{A}$ is simple,
3. a. $\underline{A}$ has no proper subalgebras,
   b. $\underline{A}$ is rigid,
   c'. $\underline{A}$ has exactly one non-trivial homomorphic image, this homomorphic image is a primal algebra.

<u>Proof:</u> („if" part) Assume $\underline{A}$ generates an arithmetical variety and the conditions a'.,b.,c. are satisfied. Since $\underline{A}$ is simple and the non-trivial uniquely determined subalgebra $\underline{B}$ of $\underline{A}$, as a primal algebra, is also simple, all subalgebras of $\underline{A}$ are simple. $\underline{A}$ is rigid and the subalgebra $\underline{B}$ of $\underline{A}$ is also rigid. Therefore all subalgebras of $\underline{A}$ have no proper automorphisms. The assumptions of 2.2-Theorem 3 are fulfilled and $\underline{A}$ is semiprimal. Since $\underline{A}$ has only one proper subalgebra, by the definition of semiprimality every function which preserves the subset $B \subseteq A$ is a term function of $\underline{A}$ and $\underline{A}$ is of the form $\underline{A}_{\underline{B}}$, i.e. $\underline{A}$ is preprimal. We can restrict our assumptions to the requirement that $\underline{B}$ is simple and rigid. Since $\underline{B}$ is the only proper subalgebra of $\underline{A}$, $\underline{B}$ has no proper subalgebras. $\underline{B}$ generates an arithmetical variety. Therefore $\underline{B}$ is primal.

Assume that $\underline{A}$ generates an arithmetical variety and that the conditions a.,b'., and c. are satisfied. The automorphism group of $\underline{A}$ is cyclical and of prime order, i.e. the automorphism group of $\underline{A}$ is generated by a permutation s of order p≠1 on A: $Aut(\underline{A}) = [s]$. s consists of cycles of the same prime length p. Now we have to show that s has no fixed elements. Let $S = \{a_0,\ldots,a_1\}$ be the set of all fixed elements of s, $S \neq \emptyset$, then $\underline{S}$ is a subalgebra of $\underline{A}$ since $f(a_0,\ldots,a_1) = f(s(a_0),\ldots,s(a_1)) = s(f(a_0,\ldots,a_1))$, i.e. $f(a_0,\ldots,a_1) \in \underline{S}$. Since $\underline{A}$ has no proper subalgebras and $S \neq \emptyset$, it holds $S = A$ and s is the identity in contradiction to p≠1. All conditions of 2.2-Theorem 6 are satisfied and $\underline{A}$ is demiprimal and of the form $\underline{A_s}$.

Assume that $\underline{A}$ generates an arithmetical variety and that the conditions a.,b.,c'. are satisfied. Then the assumptions of 2.3-Theorem 10 are fulfilled and $\underline{A}$ is hemiprimal, therefore $\underline{A}$ is of the form $\underline{A_e}$. ▲

5.2 Preprimal algebras generating either congruence distributive or congruence permutable varieties

By 5.1-Proposition 1, a preprimal algebra either generates an arithmetical variety or is constantive and simple. Exactly the preprimal algebras $\underline{A}_B$, $\underline{A}_s$, and $\underline{A}_\theta$ generate arithmetical varieties while the remaining ones are constantive and simple. It follows that, if a constantive and simple preprimal algebra generates a congruence distributive variety, it cannot generate a congruence permutable variety, and if a constantive and simple preprimal algebra generates a congruence permutable variety, it cannot generate a congruence distributive variety. If $\underline{A}$ is a preprimal algebra and $V_{\underline{A}}$ is either congruence distributive or congruence permutable, then $\underline{A}$ is constantive and simple.

Let us consider the constantive and simple preprimal algebras which generate congruence permutable varieties. Let $\underline{G} = \langle A,+\rangle$ be an elementary abelian p-group, card $A = p^m$ (p a prime number, $m \geq 1$). The term functions of the preprimal algebras $\underline{A}_\alpha$ are called affine (or quasilinear) functions with respect to $\underline{G}$. (For an affine function it holds $f(a_1,\dots,a_n) + f(b_1,\dots,b_n) - f(0,\dots,0) = f(a_1+b_1,\dots,a_n+b_n)$.)
For m=1 the term functions of $\underline{A}_\alpha$ are the linear functions investigated by Jablonskij ([1]). All constant functions c_a ($c_a(x) = a$ for all $x \in A$), the functions $x+y$, $x-y$, ax, all functions defined by

$$f(x_1,\dots,x_n) = a_0 + \sum_{i=1}^{n} \sum_{j=0}^{m} a_{ij} x_i^{p^j}$$

are affine with respect to $\underline{G}$. An algebra $\underline{A} = \langle A,F\rangle$ is said to be affine over $\underline{G}$ if all its term functions are affine with respect to $\underline{G}$. Results about the structure of affine algebras can be found in (Gumm [1]). In this paper it is proved: If $\underline{A}$ is a simple affine algebra in a congruence permutable variety, then the group $\underline{G} = \langle A,+\rangle$ is either an elementary abelian p-group or is torsion free. If $\underline{A}$ is affine over an abelian p-group, then the Malcev term p is uniquely determined by $p(x,y,z) = x-y+z$. $\underline{A}_\alpha$ is affine over $\underline{G} = \langle A,+\rangle$ and generates a congruence permutable variety. By a result of Mc Kenzie ([2]), every finite simple algebra in a congruence permutable variety is either functionally complete or affine over an elementary abelian p-group. Since preprimal algebras distinct from $\underline{A}_{\mathscr{L}}$ are not affine, the algebras $\underline{A}_{\mathscr{L}}$ are the only constantive simple preprimal algebras which generate congruence permutable varieties. Thus we obtain:

Lemma 3: Algebras $\underline{A}_{\mathscr{L}}$ are the only preprimal algebras which generate congruence permutable, but not congruence distributive varieties.

Proof: (We give a direct proof not using Mc Kenzie's theorem.) It is to prove that the Malcev term p is not a term function of a constantive simple preprimal algebra distinct from $\underline{A}_{\mathscr{L}}$. There is no order relation ϱ on A with a greatest and a least element such that $V_{\underline{A}_{\varrho}}$ is congruence permutable. Let $(x,y) \in \varrho$, $x \neq y$, then $((x,x,y),(x,y,y)) \in \varrho^3$ and $(p(x,x,y),p(x,y,y)) = (y,x) \notin \varrho$ since ϱ is an antisymmetric relation. Let τ be a totally reflexive h-ary relation, $h > 2$, and τ distinct from A^h. Then there exists an h-tuple $(a_0,\dots,a_{h-1}) \notin \tau$. From

$(a_2,a_1,a_2,a_3,a_4,\dots,a_{h-1}) \in \tau$, $(a_2,a_2,a_2,a_3,a_4,\dots,a_{h-1}) \in \tau$,

$(a_0,a_2,a_2,a_3,a_4,\dots,a_{h-1}) \in \tau$

it follows

$(p(a_2,a_2,a_0),p(a_1,a_2,a_2),p(a_2,a_2,a_2),p(a_3,a_3,a_3),\dots,p(a_{h-1},a_{h-1},a_{h-1}))$

$= (a_0,a_1,\dots,a_{h-1}) \notin \tau$.

Therefore p is not a term function of a preprimal algebra $\underline{A}_{\tau}$, $(\tau \in C)$ with $h > 2$ and is not a term function of an algebra $\underline{A}_{\delta}$, where δ is a relation on B. Let $h=2$ and let τ be a binary central relation such that $(a_0,a_1) \notin \tau$ $(\tau \neq A^2)$. Then from $(c,a_1),(c,c),(a_0,c) \in \tau$ it follows $(p(c,c,a_0),p(a_1,c,c)) = (a_0,a_1) \notin \tau$, where c is an element of the centre of the relation τ, i.e. p cannot be a term function of a preprimal algebra $\underline{A}_{\tau}$ with $h=2$. ▲

The following lemma of Rosenberg ([6]) is directly applicable to the question of congruence distributivity of a variety.

Lemma 4: Let $t_1,\dots,t_n$ be the functions of 1.3-Theorem 2. For at least one $t \in T = \{t_1,\dots,t_n\}$ the following conditions (i)-(iii) are not equivalent:

(i) $t(x,x,y) = t(x,y,x) = x$,

(ii) $t(x,y,y) = t(x,y,x) = x$,

(iii) $t = e_0$.

As an application of Lemma 4 we give a direct proof of the fact that $V_{\underline{A}_\alpha}$ is not congruence distributive. If t is affine with respect to $\underline{G}$ and (i) is satisfied, from $x+y = t(x+y,y+x,x+x) = t(x,y,x)+t(y,x,x)-t(0,0,0) = x+t(y,x,x)$ follows $t(y,x,x) = y$, i.e. (ii). If (ii) is satisfied, from $x+x = t(x+x,x+y,y+x) = t(x,x,y)+t(x,y,x)-t(0,0,0) = t(x,x,y)+x$ it follows $t(x,x,y) = x$, i.e. (i). From the identities $t(x,y,z) = t(x,0,0)+t(0,y,z) = x+t(0,y,0)+t(0,0,z) = x+t(0,0,z) = x+t(x,z,z)-t(x,0,0)+t(0,z,0) = x+x-x+0 = x$ it follows (ii) $\Longrightarrow$ (iii). Consequently, the conditions (i)-(iii) are equivalent for t quasilinear, therefore not all t_i are quasilinear and $\underline{A}_\alpha$ does not generate a congruence distributive variety. ▲

Algebras of the form $\underline{A}_\alpha$ are paraprimal algebras which have no proper subalgebras. From 5.1 it follows that the preprimal algebras $\underline{A}_B$ and $\underline{A}_s$ are also paraprimal while $\underline{A}_\theta$ are not paraprimal since they are not simple. We obtain:

<u>Theorem 5:</u> A preprimal algebra $\underline{A}$ is paraprimal if and only if one of the following mutually excluding conditions is satisfied:

1. $\underline{A}$ has no proper subalgebra,
2. $\underline{A}$ has a non-trivial automorphism,
3. $\underline{A}$ is affine with respect to $\underline{G} = \langle A,+\rangle$ (which is an elementary abelian p-group).

Now let us consider constantive simple preprimal algebras which generate congruence distributive varieties. Let C be the system of all central, totally reflexive, totally symmetric h-ary relations, $(2 \leq h \leq k)$, and let τ be a relation from C. Let $(a_0,a_1,\dots,a_{h-1}) \notin \tau$, $\tau \neq A^h$. If $h>2$, $\underline{A}_\tau$ is not 2-distributive since from $(a_0,a_1,c,\dots,c) \in \tau$, $(a_0,c,a_2,\dots,a_{h-1})$

$\in \tau$ it follows that

$(d(a_0,a_0,c),d(a_1,c,a_1),d(c,a_2,a_2),\ldots,d(c,a_{h-1},a_{h-1})) = (a_0,\ldots,a_{h-1}) \notin \tau$. Therefore d is not a term function of $\underline{A}_\tau$. We consider the function t_1 defined by $t_1(x,y,z) = \begin{cases} x \text{ if } x=z \text{ or } x=y \\ c \text{ otherwise} \end{cases}$ (c is an element of the centre). Let $\tilde{a}_0 = (a_{1o},a_{2o},a_{3o}),\ldots,\tilde{a}_{h-1} = (a_{1h-1},a_{2h-1},a_{3h-1})$ and let $(\tilde{a}_0,\ldots,\tilde{a}_{h-1}) \in \tau^3$, i.e. $(a_{1o},\ldots,a_{1h-1}) \in \tau, (a_{2o},\ldots,a_{2h-1}) \in \tau$, and $(a_{3o},\ldots,a_{3h-1}) \in \tau$. If there is an $\tilde{a}_i$ with $a_{1i} \neq a_{2i}$ and $a_{1i} \neq a_{3i}$, then $(t_1(\tilde{a}_0),\ldots,t_1(\tilde{a}_i),\ldots,t_1(\tilde{a}_{h-1})) = (t_1(\tilde{a}_0),\ldots,c,\ldots,t_1(\tilde{a}_{h-1}))$ since τ is central and totally symmetric. If for all $\tilde{a}_i$ it holds $a_{1i} = a_{2i}$ or $a_{1i} = a_{3i}$, then $(t_1(\tilde{a}_0),\ldots,t_1(\tilde{a}_{h-1})) = (a_{1o},\ldots,a_{1h-1}) \in \tau$. Therefore t_1 preserves the relation τ, i.e. t_1 is a term function of $\underline{A}_\tau$. Obviously, the function t_2 defined by $t_2(x,y,z) = \begin{cases} z \text{ if } x=y \text{ or } x=z \\ c \text{ otherwise} \end{cases}$ is also a term function of $\underline{A}_\tau$. Since all maximal classes contain the projections, the functions t_0 and t_3 defined by $t_0(x,y,z) = x$ and $t_3(x,y,z) = z$ are term functions of $\underline{A}_\tau$. t_0,t_1,t_2,t_3 satisfy the identities of 1.3-Theorem 2 and $V_{\underline{A}_\tau}$ is congruence distributive. For $h > 2$ $V_{\underline{A}_\tau}$ is 3-distributive, but not 2-distributive, and for $h = 2$ it is 2-distributive since in this case the function d defined by $d(x,x,y) = d(x,y,x) = d(y,x,x) = x$, $d(x,y,z) = c$ otherwise, (c is an element of the centre of τ), is a term function of $\underline{A}_\tau$. We obtain:

<u>Lemma 6:</u> Preprimal algebras $\underline{A}_\tau$ are 3-distributive, but not 2-distributive if $h > 2$ and they are 2-distributive if $h=2$.

The preprimal algebras $\underline{A}_\tau$ are not the only preprimal algebras which generate congruence distributive, but not congruence permutable varieties. We consider order relations ϱ with a least and a greatest element, the so-called bounded order relations. If ϱ is a lattice order, 1.3-Corollary 3 immediately gives 2-distributivity. In this case the function d defined by $d(x,y,z) = (x \wedge_\varrho y) \vee_\varrho ((x \wedge_\varrho z) \vee_\varrho (y \wedge_\varrho z))$, where $x \wedge_\varrho y$ ($x \vee_\varrho y$) denotes the infimum (supremum) of x and y with respect to ϱ, is a term function of $\underline{A}_\varrho$ and $d(x,x,y) = d(x,y,x) = d(y,x,x) = x$ are identities in

$\underset{\sim}{A}_\varrho$. Conversely we will prove that if $\underset{\sim}{A}_\varrho$ is a preprimal algebra generated by an order relation with the least element 0 and the greatest element 1 and $\underset{\sim}{A}_\varrho$ is 2-distributive, then ϱ is a lattice order on A. By assumption there is a term function d in $\underset{\sim}{A}_\varrho$ with $d(x,x,y) = d(x,y,x) = d(y,x,x) = x$. c_0 and c_1 with $c_0(x) \equiv 0$ and $c_1(x) \equiv 1$ are term functions of $\underset{\sim}{A}_\varrho$ since $\underset{\sim}{A}_\varrho$ is constantive. $d(x,y,1)$ is the supremum $x \vee_\varrho y$ of x and y with respect to ϱ and $d(x,y,0)$ is the infimum of x and y with respect to ϱ . Therefore we have

<u>Lemma 7:</u> A preprimal algebra $\underset{\sim}{A}_\varrho$ is 2-distributive if and only if ϱ is a lattice order on A.

The algebras $\underset{\sim}{A}_\varrho$ are examples for ordered algebras. By an ordered algebra we understand an algebra $\underset{\sim}{A} = \langle A,F\rangle$ such that $\langle A,\varrho\rangle$ is an ordered set and for every n-ary term function $f \in F$ it holds:

If $(x_1,y_1) \in \varrho, (i=1,\ldots,n)$, then $(f(x_1,\ldots,x_n), f(y_1,\ldots,y_n)) \in \varrho$.

The linearly ordered algebras, the lattice ordered algebras and the bounded ordered algebras are special cases of ordered algebras. In the last case ϱ is an order relation on A with the least element 0 and the greatest element 1. Lemma 7 can be generalized to certain ordered algebras:

Let $\underset{\sim}{A}$ be a bounded ordered algebra such that there are term functions c_0 and c_1 with $c_0(x) \equiv 0$, $c_1(x) \equiv 1$, then $\underset{\sim}{A}$ is 2-distributive if and only if $\underset{\sim}{A}$ is a lattice ordered algebra.

Which bounded ordered algebras are preprimal ? In this direction we will prove the following

<u>Theorem 8:</u> Let $\underset{\sim}{A} = \langle A,F\rangle$ be a bounded ordered algebra such that the following conditions are satisfied:

(i) $\underset{\sim}{A}$ is constantive,

(ii) $\underset{\sim}{A}$ is 2-distributive,

(iii) Every order homomorphism of the ordered set $\langle A,\varrho\rangle$ onto the ordered set $\langle\{0,1\},\varrho\rangle$ is a term function of $\underset{\sim}{A}$, (i.e. every function

$\varphi: A \longrightarrow \{0,1\}$ such that from $(a_1,a_2) \in \varsigma$ it followe $(\varphi(a_1),\varphi(a_2)) \in \varsigma$ is a term function).

Then $\underline{A} = \langle A,F \rangle$ is preprimal.

Proof: Since $\underline{A}$ is constantive, there are term functions c_0 and c_1 such that $c_0(x) \equiv 0$ and $c_1(x) \equiv 1$. Therefore, by (ii), $\underline{A}$ is a lattice ordered algebra and for any two elements $x,y \in A$ there exist $x \wedge_\varsigma y$ and $x \vee_\varsigma y$. All unary functions $m_b{}^a$ such that $m_b{}^a(x) = \begin{cases} a & \text{if } (b,x) \in \varsigma \\ 0 & \text{otherwise} \end{cases}$ are term functions of $\underline{A}$. If $b=0$ from $(0,x) \in \varsigma$ for all $x \in A$ by (i) we obtain that $m_b{}^a$ is a term function of $\underline{A}$. If $a=1$, we have a function $m_b{}^a: A \longrightarrow \{0,1\}$ which preserves ς and by (iii) $m_b{}^a$ is a term function of $\underline{A}$. Assume that $b \neq 0$ and $a \neq 1$. If $(b,x) \in \varsigma$, we have $m_b{}^a(x) = 1 \wedge_\varsigma a = a$ and if $(b,x) \notin \varsigma$, then it holds $m_b{}^a(x) = 0 \wedge_\varsigma a = 0$, i.e. $m_b{}^a(x) = m_b{}^1(x) \wedge_\varsigma m_0{}^a(x)$. Let $\tilde{a} = (a_1,\dots,a_n) \in A^n$. The functions $f_{\tilde{a}}$ defined by $f_{\tilde{a}}(\tilde{x}) = \begin{cases} f(\tilde{a}) & \text{if} (\tilde{a},\tilde{x}) \in \varsigma^n \\ 0 & \text{otherwise} \end{cases}$ are term functions of $\underline{A}$ since

$$f_{\tilde{a}}(x_1,\dots,x_n) = m_{a_1}{}^{f(\tilde{a})}(x_1) \wedge_\varsigma m_{a_2}{}^{f(\tilde{a})}(x_2) \wedge_\varsigma \dots \wedge_\varsigma m_{a_n}{}^{f(\tilde{a})}(x_n).$$

Every ς preserving n-ary function f has the representation

$$f(x_1,\dots,x_n) = f_{\tilde{a}_1}(x_1,\dots,x_n) \vee_\varsigma f_{\tilde{a}_2}(x_1,\dots,x_n) \vee_\varsigma \dots \vee_\varsigma f_{\tilde{a}_{(\mathrm{card}\,A)^n}}(x_1,\dots,x_n)$$

Therefore $\underline{A} = \langle A,F \rangle$ is preprimal. ▲

For an additional result on certain preprimal algebras $\underline{A}_\varsigma$ see (Quackenbush[5]). From Quackenbush's result it follows that the pseudovariety generated by an algebra $\underline{A}_\varsigma$, where ς is a bounded lattice order on A, is equivalent (as a category) to the pseudovariety of the finite bounded distributive lattices, i.e. to $HSP_f\,\underline{A}_1$, ($\underline{A}_1 = \langle \{0,1\}, \wedge, \vee, 0, 1 \rangle$, P_f: finite direct powers). (For the definition of pseudovarieties see Problems and Complements[7]).

If ς is not a lattice order relation on A, the situation is less clear. Among the orders ς given by the following diagrams, only 1 and 2 lead to algebras $\underline{A}_\varsigma$ which generate congruence distributive varieties, the remaining one does not (Knoebel[5]).

1

2

3

5.3 Preprimal algebras generating neither congruence distributive nor congruence permutable varieties

Preprimal algebras generating neither congruence distributive nor congruence permutable varieties are constantive and simple and are not of the form $\underline{A}_{\alpha}$ or $\underline{A}_{\tau}$. Which constantive and simple preprimal algebras distinct from $\underline{A}_{\alpha}$ and $\underline{A}_{\tau}$ do not generate congruence distributive varieties ? For algebras $\underline{A}_{\varsigma}$, where ς is an arbitrary bounded order relation on A, this question is open. Now we consider algebras $\underline{A}_{\delta}$, where δ is an arbitrary h-ary relation on A from B (cf. 4.1),($3 \leq h \leq$ card A). If δ is not the unit-relation (A^h), there is an h-tuple $(a_0,\dots,a_{h-1}) \notin \delta$ A function d with $d(x,z,z) = z$, $d(x,x,z) = d(x,z,x) = x$ cannot be a term function of an algebra $\underline{A}_{\delta}$ since from $(a_1,a_1,a_2,a_3,\dots,a_{h-1}) \in \delta$, $(a_0,a_1,a_1,a_3,\dots,a_{h-1}) \in \delta$ $(a_0,a_0,a_2,a_3,\dots,a_{h-1}) \in \delta$ it follows that $(d(a_1,a_0,a_0),d(a_1,a_1,a_0),d(a_2,a_1,a_2),d(a_3,a_3,a_3),\dots d(a_{h-1},a_{h-1},a_{h-1})) = (a_0,\dots,a_{h-1}) \notin \delta$. Therefore $V_{\underline{A}_{\delta}}$ is not 2-distributive.

Let Υ be a primitive h-elementary relation on the set $E_h = \{0,\dots,h-1\}$. The term functions of $\underline{A}_{\Upsilon}$ are exactly all essentially unary functions and all non-surjective functions. The functions $t_0,\dots,t_n$ from 1.3-Theorem 2, are all surjective. At least one of the t_i is essentially binary. Otherwise by $t_i(x,y,x) = x$ each t_i is either e_0 or e_2. Let j be the least index such that $t_j = e_2$ $(0 < j < n)$. From $t_i(x,x,y) = t_{i+1}(x,x,y)$ if i is even or $t_i(x,y,y) = t_{i+1}(x,y,y)$ if i is odd for z=x or z=y we obtain the contradiction $x = e_0(x,z,y) = t_{j-1}(x,z,y) = t_j(x,z,y) = y$. Thus $T = \{t_1,\dots,t_n\}$ is not included in the set of all essentially unary or non-surjective functions. It follows that the algebras $\underline{A}_{\Upsilon}$ do not generate congruence distributive varieties. In 5.5 we shall find a categorical equivalence between varieties $V_{\underline{A}_{\Upsilon}}$ and $V_{\underline{A}_{\eta}}$ (η an h-elementary relation on h^m), which preserves surjectivity. Thus homomorphic images and congruences are preserved under this equivalence. Therefore $V_{\underline{A}_{\eta}}$ is not congruence distributive. Since the congruence lattice of an algebra in $V_{\underline{A}_{\delta}}$ is isomorphic to the congruence lattice of an algebra in $V_{\underline{A}_{\eta}}$,

$D_n = A^n$, thus $D_k = A^k$ and $(0,\dots,k-1) \in D_k$, i.e. for all $a \in A$ there exists an element $c \in A$ such that $(a,c) \in D$. Together with the total symmetry and the total reflexivity of D this means, that c is in the centre of D. Hence $\underline{D}$ is central and by Theorem 9 $\underline{A}$ is preprimal.

<u>2.2 $\underline{D} \cap \underline{D}^{-1} = \Delta(\underline{A})$:</u> If $\underline{D} \cap \underline{D}^{-1} = \Delta(\underline{A})$, then $\underline{D}$ is antisymmetric. Since $D \circ D^{-1}$ is symmetric, it holds $D \neq D \circ D^{-1}$ and $D \circ D^{-1} = A^2$, ($D \circ D^{-1}$ properly contains $\Delta(\underline{A})$). We consider the algebras $\underline{D}_n \subseteq \underline{A}^n$. It follows that $D_2 = D \circ D^{-1} = A^2$ and by induction we obtain $D_k = A^k$. But then $(0,\dots,k-1) \in D_k$ and so for some $1 \in A$ $(a,1) \in D$ for all $a \in A$, i.e. 1 is a greatest element of D. From $D^{-1} \circ D = A^2$ it follows that D has a least element 0. Consider $D \circ D$. $\langle D \circ D, F \rangle$ is a subalgebra of $\underline{A}^2$. Let $a \in A$ and 1 be the greatest element of D. If $(1,a) \in D \circ D$, then for some $b \in A$ $(1,b)$ and $(b,a) \in D$. Since $(b,1) \in D$ and since D is antisymmetric, $(1,b) \in D$ means that $b=1$. Hence $(1,a) \in D$ and so $a=1$. Thus, for $a=1$, $(1,a) \in D \circ D$ and $D \circ D \neq A^2$. Since $D \circ D$ properly contains $\Delta(\underline{A})$, it follows $D \circ D = D$ and D is transitive and so is a partial order with a least and a greatest element. By Theorem 9 it follows that $\underline{A}$ is preprimal. As in 5.2 we can prove that $\underline{A}$ is a lattice ordered algebra.

<u>3. $\Delta(\underline{A}) \cap \underline{D} = \emptyset$:</u> If $\Delta(\underline{A}) \cap \underline{D} = \emptyset$, $\underline{D}$ is not reflexive. For $2 \leq n \leq k$ we consider the set $D_n = \{(a_1,\dots,a_n) \mid (a_i,b) \in D$ for some $b \in A$ and $i=1,\dots,n\}$. Then $D_2 = \{(a_1,a_2) \mid (a_1,b) \in D$ and $(a_2,b) \in D$ for some $b \in A\}$. If $D_2 \neq \emptyset$, $\underline{D}_2$ is a subalgebra of $\underline{A}^2$. D_2 is reflexive, therefore $D_2 = D$. Suppose that $D_2 = A^2$. Then by induction it would follow that $D_k = A^k$, i.e. $(0,\dots,k-1) \in D_k$ and for some $u \in A$ we would have $(a,u) \in D$ for all $a \in A$, i.e. $(u,u) \in D$ in contradiction to $\Delta(\underline{A}) \cap \underline{D} = \emptyset$. Therefore $D_2 = \Delta(A)$, i.e. $(a_1,a_2) \in D_2$ if and only if $a_1 = a_2$. From $(a_1,b) \in D$ and $(a_2,b) \in D$ it follows $a_1 = a_2$. If $D_2 = \emptyset$, there is no pair (a_1,a_2) such that for some $b \in A$ $(a_1,b) \in D$ and $(a_2,b) \in D$. Now, for all $2 \leq n \leq k$ we consider the set $D_n' = \{(a_1,\dots,a_n) \mid (b,a_i) \in D$ for some $b \in A$ and all $1 \leq i \leq n\}$. We obtain: If $(b,a_1) \in D$ and $(b,a_2) \in D$, then $a_1 = a_2$ or there is no element $b \in A$ such that $(b,a_1) \in D$ and $(b,a_2) \in D$. Thus we have proved that D is a one-to-one mapping of a subset of A onto a subset of A. The next step is the

proof that this mapping is surjective. We consider the set $A_1 = \{a \mid \text{there exists an element } b \in A \text{ such that } (b,a) \in D\}$. Let $a_1,\dots,a_n \in A_1$, i.e. $(a_1,b_1),\dots,(a_n,b_n) \in D$. Then for any term function f of $\underline{A}$ it holds: $f((a_1,b_1),\dots,(a_n,b_n)) = (f(a_1,\dots,a_n), f(b_1,\dots,b_n)) \in D$, i.e. $f(a_1,\dots,a_n) \in A_1$ and $\underline{A_1}$ is a subalgebra of $\underline{A}$. Since $\underline{A}$ has no proper subalgebra it holds $\underline{A_1} = \underline{A}$, i.e. the mapping given by D is surjective. Every element of $\underline{D}$ can be written as a pair $(a,s(a))$ where s is an automorphism which has no invariant elements and s is not the identical automorphism. Since the powers of automorphisms are again automorphisms s consists of cycles of the same prime length. By Theorem 9, $\underline{A}$ is preprimal.▲

The following picture gives a survey of the proof.

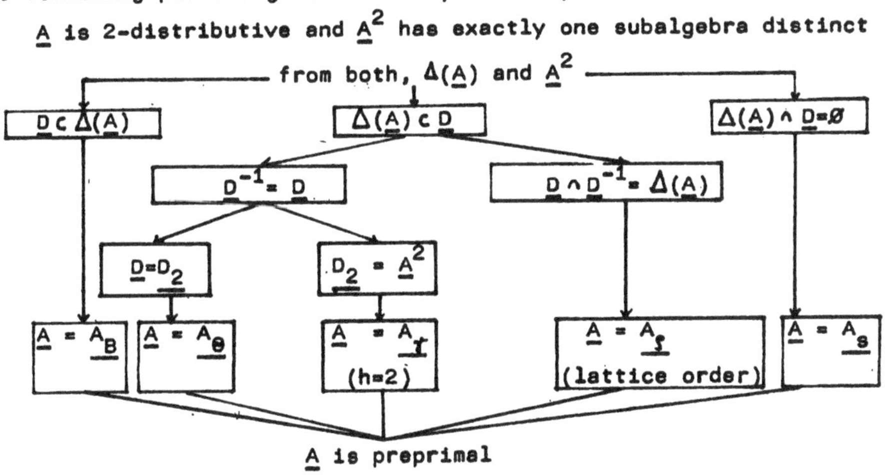

Now as a consequence of Theorem 11 we are able to state:

<u>Corollary 12:</u> Let $\underline{A} = \langle A,F\rangle$ be a finite non-trivial constantive algebra which generates a 2-distributive variety. Then $\underline{A}$ is preprimal if and only if $\underline{A}^2$ has exactly one reflexive subalgebra distinct from both, $\Delta(\underline{A})$ and $\underline{A}^2$.

An arithmetical variety is also 2-distributive since from the ternary invariant elements with cycles of the same prime length. The automorphism s corresponds to a subalgebra $\underline{D} \subset \underline{A}^2$ with $\underline{D} \cap \Delta(\underline{A}) = \emptyset$. All elements of $\underline{D}$ can be represented as pairs (a,s(a)) for all $a \in A$. The term functions of $\underline{A}$ are exactly all functions $f: A^n \longrightarrow A$ which preserve $\underline{D}$. The term functions of $\underline{A}_\tau$ (h=2) are exactly all functions which preserve a binary central relation, i.e. all functions which preserve a reflexive, symmetric, central subalgebra $\Delta(\underline{A}) \subset \underline{D} \subset \underline{A}^2$. The term functions of $\underline{A}_\varsigma$, where ς is a lattice order with a greatest and a least element, are exactly all functions which preserve a reflexive, antisymmetric, and transitive relation on A with a unit- and a zero-element such that there exist the supremum and the infimum of any two elements of A, i.e. all functions which preserve a subalgebra $\Delta(\underline{A}) \subset \underline{D} \subset \underline{A}^2$ with the corresponding properties. From Theorem 9, from the properties of the Galois correspondence Pol-Inv (Pöschel/Kalužnin[1]) and (Rosenberg[8]) it follows that in all these cases $\underline{A}^2$ has exactly one subalgebra which is distinct from both, $\Delta(\underline{A})$ and $\underline{A}^2$.

We want to discuss the converse. Let $\underline{A}$ be a finite non-trivial algebra which generates a 2-distributive variety. Assume $\underline{A}^2$ has exactly one subalgebra (up to isomorphisms) which is distinct from $\underline{A}^2$ and from $\Delta(\underline{A})$. Is $\underline{A}$ preprimal ? The answer is positive. In the proof of this proposition we shall make use of the following subsets of A^2: For a subset $D \subsetneq A^2$ let $D^{-1} = \{(a,b) \mid (b,a) \in D\}$. Let $D_1, D_2 \subsetneq A^2$, then $D_1 o D_2$ is defined by $D_1 o D_2 = \{(a,c) \mid (a,b) \in D_1$ and $(b,c) \in D_2$ for some $b \in A\}$. $D \subsetneq A^2$ is said to be symmetric if $D = D^{-1}$ and transitive if $D \circ D = D$. In the proof we apply Theorem 9 and ideas from Quackenbush's ([4]) new proof of Rosenberg's theorem (see Appendix).

Theorem 11: Let $\underline{A} = \langle A,F \rangle$ be a finite non-trivial algebra generating a 2-distributive variety. Then $\underline{A}$ is preprimal if and only if $\underline{A}^2$ has exactly one subalgebra $\underline{D}$ (up to isomorphisms) distinct from $\Delta(\underline{A})$ and from $\underline{A}^2$ ($A = \{0,\ldots,k-1\}$). (In this way one obtains the preprimal algebras $\underline{A}_B$, $\underline{A}_\theta$, $\underline{A}_\tau$(h=2), $\underline{A}_\varsigma$(ς lattice order), and $\underline{A}_s$.)

Proof (of the „if" part): For a subalgebra $\underline{D} \subset \underline{A}^2, \underline{D} \neq \Delta(\underline{A})$ there are exactly the following cases: 1. $\underline{D} \subset \Delta(\underline{A})$, 2. $\Delta(\underline{A}) \subset \underline{D}$, 3. $\Delta(\underline{A}) \cap \underline{D} = \emptyset$. $\Delta(\underline{A}) \cap \underline{D} \neq \emptyset, \neq \underline{D}, \neq \Delta(\underline{A})$ is impossible since, otherwise, we would have another subalgebra of $\underline{A}^2$ which is distinct from $\Delta(\underline{A})$, $\underline{A}^2$, and $\underline{D}$.

Let us discuss the three cases:

1. $\underline{D} \subset \Delta(\underline{A})$: By Theorem 9 every function which preserves the subalgebra $\underline{D}$ of $\underline{A}^2$ is a term function of $\underline{A}$. The subalgebras $\underline{D} \subset \Delta(\underline{A})$ correspond one-to-one to subalgebras $\underline{B} \subset \underline{A}$; therefore every function which preserves $\underline{D}$ is a term function of $\underline{A}$ and $\underline{A}$ is preprimal.

2. $\Delta(\underline{A}) \subset \underline{D}$: $\underline{D}^{-1}$ is a subalgebra of $\underline{A}^2$ and $\Delta(\underline{A}) \subset \underline{D}^{-1}$. There are the following possibilities: 2.1 $\underline{D}^{-1} = \underline{D}$, 2.2 $\underline{D}^{-1} \neq \underline{D}$, but $\underline{D}^{-1}$ is isomorphic to $\underline{D}$ and it holds $\underline{D} \cap \underline{D}^{-1} = \Delta(\underline{A})$ since, if $\underline{D} \cap \underline{D}^{-1} \supset \Delta(\underline{A})$, then $\underline{D} \cap \underline{D}^{-1}$ is a subalgebra of $\underline{A}^2$ which is distinct from $\underline{D}$, $\Delta(\underline{A})$, and from $\underline{A}^2$.

We consider the first case:

2.1 $\underline{D}^{-1} = \underline{D}$: If $\underline{D}^{-1} = \underline{D}$, then $\underline{D}$ is symmetric. For $2 \leq n \leq \text{card } A = k$ let $D_n = \{(a_1,\dots,a_n) \mid \text{for some } b \in A, (a_i,b) \in D \text{ for all } 1 \leq i \leq n\}$.

$\langle D_n, F\rangle$ is a totally symmetric subalgebra of $\underline{A}^n$. For D_2 it follows: either $D = D_2$ or $D_2 = A^2$. If $(a_1,a_2) \in D$, then choose $a=a_1$. Since $(a_1,a_1) \in D$ and $(a_2,a_1) \in D$ it follows $(a_1,a_2) \in D_2$, i.e. $D \subseteq D_2 \subseteq A^2$. By our assumption, either $D = D_2$ or $D_2 = A^2$.

2.1.1 $\underline{D} = \underline{D}_2$ $(\underline{D}_2 \neq \underline{A}^2)$: If $(a_1,a_3) \in D$ and $(a_2,a_3) \in D$, then $(a_1,a_2) \in D$. Since D is symmetric, this means that D is transitive, i.e. $D \circ D = D$. Hence, D is a non-trivial equivalence relation,(distinct from the zero- and the unit-equivalence). By Theorem 9 all functions $f: A^n \to A$ which preserve D are term functions of $\underline{A}$ and $\underline{A}$ is preprimal.

2.1.2 $\underline{D}_2 = \underline{A}^2$ $(\underline{D} \neq \underline{D}_2)$: We prove by induction that $D_n = A^n$ for all $2 \leq n \leq k$. If $D_{n-1} = A^{n-1}$, then D_n is totally reflexive. Since d with $d(x,x,y) = d(x,y,x) = d(y,x,x) = x$ is a term function of $\underline{A}$, these are identities in $\underline{A}^n$. For all $a_0,\dots,a_{n-1}$ from $(a_1,a_1,a_2,a_3,\dots,a_{n-1}) \in D_n$, $(a_0,a_1,a_1,a_3,\dots,a_{n-1}) \in D_n$, $(a_0,a_0,a_2,a_3,\dots,a_{n-1}) \in D_n$ it follows that $(d(a_1,a_0,a_0),d(a_1,a_1,a_0),d(a_2,a_1,a_2),d(a_3,a_3,a_3),\dots,d(a_{n-1},a_{n-1},a_{n-1}))$ $= (a_0,\dots,a_{n-1}) \in D_n$ (for all $n > 2$), i.e. $A^n \subseteq D_n$. Therefore we obtain

(5.5), $V_{\underline{A}_\delta}$ is also not congruence distributive.

We remark that Rosenberg has given a direct proof of this fact using the notion of a wreath algebra (Rosenberg[6],[7]).

The results of 5.1, 5.2, 5.3 are summarized in a table:

variety	congruence distributive	congruence permutable
$V_{\underline{A}_\rho}$	?	-
$V_{\underline{A}_s}$	+	+
$V_{\underline{A}_\alpha}$	-	+
$V_{\underline{A}_\theta}$	+	+
$V_{\underline{A}_B}$	+	+
$V_{\underline{A}_\tau}$	+	-
$V_{\underline{A}_\delta}$	-	-

5.4 Algebraic characterization of preprimal algebras generating 2-distributive varieties

Recall that by definition a variety V is 2-distributive if there is a term $\underline{d}$ such that $\underline{d}(x,x,y) = \underline{d}(x,y,x) = \underline{d}(y,x,x) = x$ are identities in V. A finite non-trivial algebra $\underline{A}$ is said to be 2-distributive if $\underline{A}$ generates a 2-distributive variety (1.3-Theorem 2 and 1.3-Corollary 3). The following Theorem 9 is a consequence of a more general theorem of Baker/Pixley ([1]).

Theorem 9: Let $\underline{A}$ be a finite non-trivial algebra which generates a 2-distributive variety. Then for any positive integer n an n-ary function $f: A^n \longrightarrow A$ is a term function of $\underline{A}$ if and only if f preserves every subalgebra of $\underline{A}^2 = \underline{A} \times \underline{A}$ (1.2).

For a primal algebra we obtain the following well-known fact:

Corollary 10: Let $\underline{A}$ be a finite non-trivial algebra which generates a 2-distributive variety. Then $\underline{A}$ is primal if and only if $\underline{A}^2$ has only two subalgebras, the diagonal $\Delta(\underline{A})$ and $\underline{A}^2$.

Now let us consider preprimal algebras which generate 2-distributive varieties. By 5.1, 5.2, and 5.3 exactly the preprimal algebras $\underline{A_B}$, $\underline{A_\theta}$, $\underline{A_s}$, $\underline{A_\tau}$, (if τ is a binary central relation), and $\underline{A_\varsigma}$, (if ς is a lattice order with a greatest and a least element) are 2-distributive. The subalgebras B with $\emptyset \neq \underline{B} \subset \underline{A}$ correspond one-to-one to subalgebras $\underline{D} \subset \Delta(\underline{A}) \subset \underline{A}^2$. The term functions of $\underline{A_B}$ are exactly all functions $f: A^n \longrightarrow A$ which preserve the subalgebra $\emptyset \neq \underline{B} \subset \underline{A}$, i.e. all functions which preserve the subalgebra $\underline{D} \subset \Delta(\underline{A}) \subset \underline{A}^2$. The term functions of $\underline{A_\theta}$ are exactly all functions $f: A^n \longrightarrow A$ which preserve a non-trivial equivalence relation θ, i.e. all functions which preserve a reflexive symmetric and transitive subalgebra $\Delta(\underline{A}) \subset \underline{D} \subset \underline{A}^2$. The term functions of $\underline{A_s}$ are exactly all functions $f: A^n \longrightarrow A$ which preserve an automorphism s of A without discriminator t by $t(x,y,z) = t(x,t(x,y,z),z)$ we obtain the majority term function d. This leads to a further characterization of preprimal algebras which generate arithmetical varieties. At first we ask: What common property does the only subalgebra $\underline{D}$, ($\underline{D} \neq \Delta(\underline{A})$, $\underline{D} \neq \underline{A}^2$) in the cases $\underline{A}_B$, $\underline{A}_\theta$, and $\underline{A}_s$ have ? For the answer we need the concept of a rectangular subalgebra $\underline{D}$ of $\underline{A}^2$ (Pixley[7]). A subalgebra $\underline{D}$ of $\underline{A}^2$ is said to be rectangular if for all $x,u,y,v \in A$ from $(x,y),(x,v),(u,v) \in \underline{D}$ it follows $(u,y) \in \underline{D}$. A geometrical interpretation is that $\underline{D}$ has to contain the fourth vertex of a rectangle whenever it contains any three of the vertices.

Let $\underline{D}$ be a reflexive, symmetric, and transitive subalgebra of $\underline{A}^2$ (case 2.1.2), and let $(x,y),(x,v),(u,v) \in D$, then $(y,x) \in D$ and from $(y,x) \in D$, $(x,v) \in D$ it follows $(y,v) \in D$. From $(y,v) \in D$ and $(v,u) \in D$ it follows that $(y,u) \in D$ and $(u,y) \in D$, i.e. $\underline{D}$ is a rectangular subalgebra of $\underline{A}^2$. Conversely, let $\underline{D}$ be a reflexive rectangular subalgebra of $\underline{A}^2$: $\Delta(\underline{A}) \subset \underline{D} \subset \underline{A}^2$. From $(x,y),(x,x),(y,x) \in \underline{D}$ it follows $(y,x) \in \underline{D}$, i.e. $\underline{D}$ is symmetric and from $(x,y),(y,z),(y,y) \in \underline{D}$ it follows $(x,z) \in D$, i.e. $\underline{D}$ is transitive. Therefore $\underline{D}$ is a non-trivial equivalence relation. The subalgebras $\underline{D}$ of $\underline{A}^2$ in the cases 2.1.2 and 2.2 of the proof are not rectangular. The subalgebra $\underline{D} \subset \Delta(\underline{A})$ in the first case is rectangular. In the case 3. from $(x,y),(x,v),(u,v) \in D$ it follows $y=v$, thus $(u,y) = (x,y) \in \underline{D}$. We obtain:

<u>Corollary 13:</u> Let $\underline{A} = \langle A,F \rangle$ be a finite non-trivial algebra which generates an arithmetical variety. Then $\underline{A}$ is preprimal if and only if $\underline{A}^2$ has exactly one rectangular subalgebra which is distinct from $\underline{A}^2$ and from $\Delta(\underline{A})$.

5.5 Subdirectly irreducible elements, representation theorems

Now here are the general methods of the determination of all subdirectly irreducible algebras in a variety generated by a preprimal algebra $\underline{A}$. If $V_{\underline{A}}$ is a congruence distributive variety, by 1.4-Theorem 6, the subdirectly irreducible algebras of $V_{\underline{A}}$ are homomorphic images of subalgebras of $\underline{A}$. In the case of paraprimal algebras we apply the results of Clark/Krause ([1] , [2]). In the other cases we use the fact that a nontrivial algebra $\underline{A}$ is subdirectly irreducible if and only if the congruence lattice of $\underline{A}$ has one and only one atom which is contained in every congruence distinct from the zero-congruence.

With the knowledge of all subdirectly irreducible algebras of $V_{\underline{A}}$ we are able to give a representation theorem of $V_{\underline{A}}$. By 4.2-Corollary 4, every preprimal algebra which is not of the form $\underline{A_\theta}$, is simple. $\underline{A_\theta}$ is subdirectly irreducible since it has exactly one non-trivial congruence relation. We obtain:

Proposition 14: Every preprimal algebra is subdirectly irreducible.

Now we shall characterize all subdirectly irreducible algebras in a variety $V_{\underline{A}}$ generated by a preprimal algebra $\underline{A}$. Preprimal algebras of the forms $\underline{A_\tau}$ and $\underline{A_s}$ have the following properties: They generate congruence distributive varieties, they are simple and have no proper subalgebras. Therefore by 1.4-Corollary 7, it holds: $V_{\underline{A_\tau}} = IP_s\underline{A_\tau}$ and $V_{\underline{A_s}} = IP_s\underline{A_s}$. If $B = \{b\}$, respectively ϱ is an order relation on A with a greatest and a least element such that $V_{\underline{A_\varrho}}$ is congruence distributive, it holds: $V_{\underline{A_{\{b\}}}} = IP_s\underline{A_{\{b\}}}$ and $V_{\underline{A_\varrho}} = \overline{IP}_s\underline{A_\varrho}$, i.e. preprimal algebras $\underline{A_\tau}, \underline{A_s}, \underline{A_{\{b\}}}, \underline{A_\varrho}$ ($V_{\underline{A_\varrho}}$ congruence distributive) are categorical. There are no subdirectly irreducible elements in these varieties distinct from the generating algebras. By 1.4-Theorem 6 and by 5.1-Theorem 2 for preprimal algebras of the form $\underline{A_B}$ (card $B > 1$) it holds $V_{\underline{A_B}} = IP_sS\underline{A_B} = IP_s\{\underline{A_B}, \underline{B}\}$. $\underline{B}$ is the uniquely determined primal subalgebra of $\underline{A_B}$. $\underline{A_B}$ and $\underline{B}$ are the only subdirectly irreducible algebras in $V_{\underline{A_B}}$. $\underline{A_B}$ is an example for a semicategorical algebra. For preprimal algebras $\underline{A_\theta}$ it holds $V_{\underline{A_\theta}} = IP_sH\underline{A_\theta} = IP_s\{\underline{A_\theta},\underline{A_\theta/\theta}\}$. $\underline{A_{\theta/\theta}}$ is the uniquely determined non-trivial homomorphic image of $\underline{A_\theta}$. $\underline{A_\theta}$ is a primal algebra, therefore it is subdirectly irreducible and $V_{\underline{A_\theta}}$ has no other subdirectly irreducible algebras. $\underline{A_\theta}$ is an example for a hemicategorical algebra. Algebras of the form $\underline{A_\alpha}$ are plain and paraprimal. By 1.5-Theorem 9, each algebra of $V_{\underline{A_\alpha}}$ is isomorphic to a subdirect product of $\underline{A_\alpha}$ and $\underline{A_\alpha^*}$: $V_{\underline{A_\alpha}} = IP_s\{\underline{A_\alpha},\underline{A_\alpha^*}\}$. Here it holds $\underline{A_\alpha^*} = \underline{A_\alpha}^2/\theta$. For $A = \{0,\ldots,k-1\}$ θ is the congruence relation defined on $\underline{A_\alpha}^2$ by the partition $\{(0,0),(1,1),\ldots,(k-1,k-1)\}$ $\{(0,1),(1,2),\ldots,(k-1,0)\}$ $\{(0,2),(1,3),\ldots,(k-2,0)\},\ldots,$ $\{(0,k-1),(1,0),\ldots,(k-1,k-2)\}$. θ is uniquely determined by the diagonal and the other congruence classes contain also k elements. $\underline{A_\alpha^*}$ has a one-element subalgebra, is plain, paraprimal, and categorical. There are no subdirectly irreducible algebras in $V_{\underline{A_\alpha}}$ distinct from $\underline{A_\alpha}$ and $\underline{A_\alpha^*}$.

The algebras generated by relations from B are considerably more complicate than the algebras $\underline{A_s}$, $\underline{A_B}$, $\underline{A_\theta}$, $\underline{A_\gamma}$, or $\underline{A_\alpha}$. We describe the subdirectly irreducible algebras of $V_{\underline{A_\delta}}$ in three steps. This development is given by Knoebel ([5]). At first the subdirectly irreducible algebras from $V_{\underline{A_\iota}}$ (ι a primitive h-elementary relation) are described. In the second step we establish a categorical equivalence between the equational classes $V_{\underline{A_\iota}}$ and $V_{\underline{A_\eta}}$ which preserves surjective homomorphisms. In the last step we shall find an isomorphism between the congruence lattice of any algebra of $V_{\underline{A_\eta}}$ and the congruence lattice of a certain algebra in $V_{\underline{A_\delta}}$. From this it follows a one-to-one correspondence between the subdirectly irreducible algebras of the three varieties.

Let ι be a primitive h-elementary relation on the set $E_h = \{0,\ldots,h-1\}$, i.e. $(a_0,\ldots,a_{h-1}) \in \iota$ if and only if at least two of the a_i $(i=0,\ldots,h-1)$ are equal. Every essential or non-surjective function on A is a term function of $\underline{A_\iota}$. There are two classes of term functions of $\underline{A_\iota}$:

1. If f is non-surjective, we call it flat.

2. If f is surjective and depends on only one argument and so is a permutation on that one argument, call it sharp.

Since $\underline{A}_I$ generates $V_{\underline{A}_I}$, the terms flat and sharp extend to term functions in any algebra of the variety. Since $\underline{A}_I$ is constantive, $c_o,\ldots,c_{k-1}$ with $c_o(x)\equiv 0,\ldots,c_{k-1}(x)\equiv k-1$ are term functions of $\underline{A}_I(A = \{0,\ldots,k-1\}$. We shall show that $\underline{A}_I$ is the only subdirectly irreducible algebra in $V_{\underline{A}_I}$ for which $0 \neq 1$. Let $\underline{B} \in V_{\underline{A}_I}$ and $0 \neq 1$ in $\underline{B}$ and $\underline{B}$ be not isomorphic to $\underline{A}_I$. We have to determine two non-trivial congruences θ_1 and θ_2 on $\underline{B}$ such that $\theta_1 \wedge \theta_2$ is the zero-congruence.

Lemma 15: Let $\underline{A} = \langle\{0,\ldots,k-1\},F\rangle$ be a finite non-trivial algebra and let every function on A with a range of no more than two elements be a term function of $\underline{A}$. Let $\underline{B} \in V_{\underline{A}}$. We define a binary relation θ_2 on $\underline{B}$ by $(x,y) \in \theta_2$ iff $x \S y = 1 \in \mathrm{Id}\,\underline{B}$ with $x \S y = \begin{cases} 1 \text{ if } x=y \\ 0 \text{ otherwise.}\end{cases}$

If there is an element $b \in B$ such that b is not congruent to one of the elements $0^{\underline{B}},\ldots,(k-1)^{\underline{B}}$ by θ_2, then there are non-trivial congruences σ, τ of $\underline{B}$ such that $\sigma \wedge \tau = \theta_2$.

Proof: The following functions are term functions of $\underline{A}$:

$x \vee y = \begin{cases} 0 \text{ if } x=y=0 \\ 1 \text{ otherwise}\end{cases}$, $x \wedge y = \begin{cases} 1 \text{ if } x=y=1 \\ 0 \text{ otherwise}\end{cases}$, $x \supset y = Nx \vee y$.

All constants in $\underline{B}$ are distinct. Suppose there are two constants $c,d \in A$ such that $c \neq d$ in $\underline{A}$ but $c=d \in \mathrm{Id}\,\underline{B}$. Every function on A with a range of no more than two elements is a term function, any two elements of $\underline{A}$ may be taken into any other two by some term function. Thus any two constants would have to be equal to, say 0. In $\underline{A}$ we have the following identities: $x \vee x = x$, $(x \S 0) \vee (x \S 1) \vee \ldots \vee (x \S (k-1)) = 1$. In $\underline{B}$ is $b \S 0 = (b \S 0) \vee (b \S 0) \vee \ldots \vee (b \S 0) = (b \S 0) \vee (b \S 1) \vee \ldots \vee (b \S (k-1)) = 1$. Thus $(b,0) \in \theta_2$ in contradiction to the assumption. Therefore all constants in $\underline{B}$ are distinct. For some $c \in A$ it must be that $b \S c^{\underline{B}} \neq 0$ in $\underline{B}$. Suppose otherwise, then by $(x \S 0) \vee (x \S 1) \vee \ldots \vee (x \S (k-1)) = 1$, which holds also in $\underline{B}$, $0 = 1$. Let $s = b \S c^{\underline{B}}$ and $t = Ns$. It holds $s \neq 1$, otherwise $b \S c^{\underline{B}} = s = 1$ and thus $(b,c^{\underline{B}}) \in \theta_2$, which is a contradiction. Further it holds $t \neq 0$ since $Nt = NNs = s \neq 1$; similarly $t \neq 1$. For s and t it holds $s \vee t = 1$ and $s \wedge t = 0$. We define two congruences σ, τ on $\underline{B}$ by $(x,y) \in \sigma$ iff $s \supset (x \S y) = 1$, $(x,y) \in \tau$ iff $t \supset (x \S y) = 1$. σ is non-trivial since from $s = Nt \in Id\ \underline{B}$, $x \S Nx = 0 \in Id\ \underline{B}$, and $x \supset 0 = Nx \in Id\ \underline{B}$, it follows in $\underline{B}$ that $s \supset (s \S t) = s \supset 0 = t \neq 1$; hence $(s,t) \notin \sigma$. The reflexivity and symmetry of σ follow from the following identities in $\underline{A}$: $x \S x = 1$, $x \supset 1 = 1$, $x \S y = y \S x$. Let $(x,y) \in \sigma$ and $(y,z) \in \sigma$, i.e. $s \supset (x \S y) = 1 = s \supset (y \S z)$. From the identities $(x \supset z) \vee ((x \supset y) \wedge (y \supset z)) = (x \supset z) \in Id\ \underline{B}$, $(x \supset y) \wedge (x \supset z) = x \supset (y \wedge z) \in Id\ \underline{B}$, $((x \S y) \wedge (y \S z)) \supset (x \S z) = 1 \in Id\ \underline{B}$ we obtain $s \supset ((x \S y) \wedge (y \S z)) = 1$. In $\underline{B}$ it holds: If $x \supset y = 1$ and $y \supset z = 1$, then $x \supset z = 1$. We obtain $s \supset (x \S z) = 1$, i.e. $(x,z) \in \sigma$, i.e. transitivity. Let f be any term function of $\underline{B}$ and (x_0,y_0) $(x_1,y_1),\ldots \in \sigma$, i.e. $s \supset (x_0 \S y_0) = 1$, $s \supset (x_1 \S y_1) = 1, \ldots$. It follows $(s \supset (x_0 \S y_0)) \wedge (s \supset (x_1 \S y_1)) \wedge \ldots = 1$ and $s \supset ((x_0 \S y_0) \wedge (x_1 \S y_1) \wedge \ldots) = 1$. From $((x_0 \S y_0) \wedge (x_1 \S y_1) \wedge \ldots) \supset (f(x_0,\ldots) \S f(y_0,\ldots)) = 1$ and $s \supset ((x_0 \S y_0) \wedge (x_1 \S y_1) \wedge \ldots) = 1$ it follows $((f(x_0,x_1,\ldots), f(y_0,y_1,\ldots)) \in \sigma$, i.e. the congruence property. In the same manner it can be proved that τ is a congruence. Finally we show that $\sigma \cap \tau = \theta_2$. $(w \supset x) \wedge (Nw \supset x) = x$ is an identity in $\underline{A}$. Let $(x,y) \in \sigma \cap \tau$, i.e. $s \supset (x \S y) = 1$ and $t \supset (x \S y) = 1$. Therefore by the above identity (and $t = Ns$) we have $x \S y = 1$ and hence $(x,y) \in \theta_2$. ▲

To define the congruence relation θ_1, we need the notion of a flat element of $\underline{B}$: b is flat if there is a flat unary term function of $\underline{B}$ such that $f(b) = b$. Note that for any $b_0,\ldots, \in \underline{B}$ and any flat term function f the element $f(b_0,b_1,\ldots,)$ is flat. Now we define θ_1 by $(x,y) \in \theta_1$ iff $x=y$ or x and y are both flat elements. θ_1 is an equivalence relation and it is shown to have the congruence property as follows. If f is a flat term function of $\underline{B}$ and $(x_0,y_0),(x_1,y_1),\ldots \in \theta_1$, then it follows $(f(x_0,x_1,\ldots),f(y_0,y_1,\ldots)) \in \theta_1$ since $f(x_0,x_1,\ldots)$ and $f(y_0,y_1,\ldots)$ are both flat elements. If f is a sharp term function of $\underline{B}$ and $(x,y) \in \theta_1$, then there are two cases; when x and y are flat, then f(x) and f(y) are also flat since a composition of both sharp and flat term functions is flat, and hence $(s(x),s(y)) \in \Theta_1$; otherwise $x = y$. (For instance 0 and 1 are flat elements and distinct; therefore Θ_1 is non-trivial.) If $\sigma \wedge \tau$ is the zero-congruence, $\underline{B}$ is subdirectly irreducible. Otherwise, we prove that $\Theta_1 \wedge \Theta_2$ is the zero-congruence. Let $(x,y) \in \Theta_1 \wedge \Theta_2$, then it holds $x \S y = 1$ and we can assume that x and y are flat, i.e. $f(x) = x$, $g(y) = y$. If x and y are constants, then $x = y$ since $x \S y = 1$, therefore we can assume that one is not a constant. Let rng f be the range of f in $\underline{A_I}$. We can assume that $0 \in$ rng f and $0 \in$ rng g, since otherwise we can permute x and y by some sharp term function s; for there is the identity $s(x) \S s(y) = x \S y$. We consider the following binary term function $\wedge_C$ of $\underline{A_I}$: $x \wedge_C y = \begin{cases} y & \text{if } x=1 \text{ and } y \in C \\ 0 & \text{otherwise} \end{cases}$ where C is a subset of the carrier $\{0,\dots,h-1\}$ of $\underline{A_I}$. In $\underline{A_I}$ it holds the following identity: $(f(x) \S g(y)) \wedge_{\mathrm{rng}\, f} f(x) = (f(x) \S g(y)) \wedge_{\mathrm{rng}\, g} g(y)$. From the following identities in $\underline{A_I}$ $1 \wedge_{\mathrm{rng}\, f} f(x) = f(x)$, $1 \wedge_{\mathrm{rng}\, g} g(y) = g(y)$ and the fact that $f(x) \S g(y) = x \S y = 1$ in $\underline{B}$ it follows that $x = y$. This shows that, up to isomorphism, $\underline{A_I}$ is the only subdirectly irreducible algebra in $V_{\underline{A_I}}$ for which $0 \neq 1$.

Now we assume that $0=1$ in each algebra of $V_{\underline{A_I}}$ considered in the following. Since the sharp term functions are two-fold transitive on A, all the constants are equal in this class of algebras, say equal to β.

<u>Lemma 16:</u> If f is a flat term function of $\underline{A_I}$, then $f = \beta \in \mathrm{Id}\,\underline{B}$. If s is a sharp term function of $\underline{A_I}$, then s is also a sharp term function of $\underline{B}$.

<u>Proof:</u> Without loss of generality, assume that f is flat with values in $E_{h-1} = \{0,1,\dots,h-2\}$. For instance, if f is a binary term function, f can be written as $f(x,y) = \bigvee_{i \le h-2} \bigvee_{j \le h-2} f(i,j) \wedge \delta_i(x) \wedge \delta_j(x)$. Here $\bigvee$ is iterated $\vee$, and $\vee, \wedge$ and δ_i are the usual maximum, minimum, and Kronecker δ functions on A whose range is restricted to the set $\{0,\dots,h-2\}$ in the following manner: if the value should originally be h-1, it is changed to h-2. Therefore, these functions are flat term functions. The f(i,j) are always constants and hence equal β in $\underline{B}$, thus $f=\beta$. The sharp term functions of $\underline{A_I}$ are permutations and satisfy the identities

$s(s^{-1}(x)) = x = s^{-1}(s(x))$. These identities also hold in $\underline{B}$, and therefore $s^{\underline{B}}$ is a permutation. In $\underline{B}$ we have $s(\beta) = \beta$ and $s(x) \neq x$ for $x \neq \beta$ (otherwise $x = s^{-1}(s(x)) = \beta$).▲

For every $\underline{B} \in V_{\underline{A}_I}$ with $0 = 1$ we define the permutation group $\underline{B \setminus \{\beta\}} = \langle B\setminus\{\beta\}, \overline{P} \rangle$, where $P = \{ s/B\setminus\{\beta\} \mid s$ is a sharp term function of $\underline{B}\}$. We consider the free algebra $\underline{A_0}$ on one generator in the class of all algebras of $V_{\underline{A}_I}$ with $0 = 1$. By Lemma 15 and by Lemma 16 its carrier is $S_h \cup \{\beta\}$ ($\underline{S_h}$ is the symmetric group on $\{0,\ldots,h-1\}$). The term functions of A_0 are defined by

$$f(a_0,\ldots,a_{n-1}) = \begin{cases} \beta \text{ if } f \text{ is flat} \\ \beta \text{ if } f \text{ is sharp at } i \text{ and } a_i = \beta \\ s(a_i) \text{ if } f \text{ is sharp at } i \text{ with permutation } s \text{ and } a_i \neq \beta. \end{cases}$$

The permutation group corresponding to $\underline{A_0}$ is $\underline{A_0 \setminus \{\beta\}} = \langle \underline{S_h}, \underline{S_h} \rangle$. A congruence on $\underline{A_0}$ is uniquely determined by a congruence of $\underline{A_0 \setminus \{\beta\}}$ and is the partition of all left cosets of a subgroup $\underline{H}$ of $\underline{S_h}$ together with β. A congruence on $\underline{A_0 \setminus \{\beta\}}$ is uniquely determined by the subgroup $\underline{H}$ of $\underline{S_h}$ of permutations related to the identity; left multiplication gives the action of the group on the left cosets of $\underline{H}$. For two non-trivial congruences on $\underline{A}$ the meet is the zero-congruence if and only if the meet of the corresponding subgroups is the one-element subgroup of $\underline{S_h}$. Therefore $\underline{A_0}$ is subdirectly irreducible. If $\underline{B\setminus\{\beta\}}$, as a permutation group, is transitive, $\underline{B}$ has one generator; since $\underline{A_0}$ is the free algebra on one generator, any algebra $\underline{B} \in V_{\underline{A}_I}$ with $0 = 1$ and one generator must be a homomorphic image of $\underline{A_0}$. $\underline{B} \in V_{\underline{A}_I}$ with $0 = 1$ and one generator is subdirectly irreducible if there are two non-trivial subgroups of $\underline{B\setminus\{\beta\}}$ and the meet of these subgroups is the one-element subgroup. Conjugate subgroups lead to isomorphic subdirectly irreducible algebras. If $\underline{B\setminus\{\beta\}}$ is intransitive, then $\underline{B}$ is reducible. A system of intransitivity is a proper partition of $\underline{B\setminus\{\beta\}}$ such that any permutation takes an element of a block of the partition into the same block. It suffices to consider a system with only two blocks, i.e. B is a disjoint union of the blocks E and D. Let $\underline{E_1}$ and $\underline{D_1}$ be the corresponding algebras in $V_{\underline{A}_I}$. The carrier of $\underline{E_1}$ is $E \cup \{ß\}$, the flat term functions are interpreted in $\underline{E_1}$ as ß; and for any sharp term function e it holds

$$e^{\underline{E_1}}(x) = \begin{cases} ß \text{ if } x = ß \\ e^{\underline{B}}(x) \text{ if } x \in E \end{cases}$$

and similarly for $\underline{D_1}$. Then $\underline{B}$ is isomorphic to a subalgebra of $\underline{E_1} \times \underline{D_1}$ by φ and $\varphi(e) = (e,ß)$, $\varphi(d) = (ß,d)$, $\varphi(ß) = (ß,ß)$, $(e \in E, d \in D)$. $\underline{E_1}$ and $\underline{D_1}$ are non-trivial and therefore $\underline{B}$ is reducible. Then we obtain the following theorem:

<u>Theorem 17:</u> In $V_{\underline{A_\rceil}}$ there are finitely many subdirectly irreducible algebras, all of them are finite.

Let $\rceil$ be the primitive h-elementary relation on E_h, as before, and let η be the h-elementary relation on E_h^m which is the m-fold power of $\rceil$ i.e. $((a_{00},a_{01},\ldots,a_{0m-1}),(a_{10},a_{11},\ldots,a_{1m-1}),\ldots,(a_{h-10},a_{h-11},\ldots)) \in \eta$ if and only if for every $j < m$ at least two of the elements $a_{0j},\ldots,a_{h-1j}$ are equal. Let $\underline{A_\rceil}$ and $\underline{A_\eta}$ be the corresponding algebras and $V_{\underline{A_\rceil}}$, $V_{\underline{A_\eta}}$ the varieties generated by them. The subdirectly irreducible algebras of $V_{\underline{A_\eta}}$ are determined by finding a categorical equivalence between $V_{\underline{A_\rceil}}$ and $V_{\underline{A_\eta}}$ which preserves surjectivity. Thus subdirect products and homomorphic images are preserved by this equivalence, therefore there is a one-to-one correspondence between the isomorphism classes of irreducible algebras in the two varieties. We apply the following lemma:

<u>Lemma 18:</u> Assume $\varrho \subseteq A^h$, where $h \geq 1$. Let $\underline{A} = \underline{A_\varrho}$ and $B = A^m, m > 1$. Let $\underline{B} = \underline{A_{\varrho^{(m)}}}$ where $\varrho^{(m)} \subseteq B^h$ and $\varrho^{(m)}$ is the m-fold power of ϱ. Then it holds

(+) f is a n-ary term function of $\underline{B}$ iff for all $i < m$ f_i is a mn-ary term function of $\underline{A}$. Here $f_i: A^{mn} \to A$ is the ith projection of f, i.e. f_i is the composition of e_i and f. The diagram

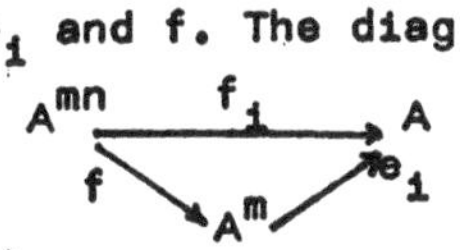

is commutative. The converse is also true: If $\underline{A} = \underline{A_\varrho}$, $B = A^m$ and (+) hold, then the set of all term functions of $\underline{B}$ is equal to the set of all term functions of $A_{\underline{f}(m)}$.

Lemma 19: If two algebras $\underline{A}_o$ and $\underline{B}_o$ satisfy the condition (+), if $V_{\underline{A}_o}$ and $V_{\underline{B}_o}$ are the varieties generated by $\underline{A}_o$ and $\underline{B}_o$, and if $\mathfrak{A}$ and $\mathfrak{B}$ are the corresponding categories, then $\mathfrak{A}$ is categorically equivalent to $\mathfrak{B}$ and surjectivity is preserved.

Proof: We apply the following characterization of the categorical equivalence: Two categories $\mathfrak{A}$ and $\mathfrak{B}$ are categorically equivalent iff there is a functor $F: \mathfrak{A} \longrightarrow \mathfrak{B}$ and an isomorphism-preserving functor G from the class $V_{\underline{B}}$ of objects of $\mathfrak{B}$ to the class $V_{\underline{A}}$ of objects of $\mathfrak{A}$ such that

(i) $G(F(\underline{A})) \cong \underline{A}$ for all $A \in V_{\underline{A}}$

(ii) $F(G(\underline{B})) \cong \underline{B}$ for all $\underline{B} \in V_{\underline{B}}$.

Let $\underline{A} \in V_{\underline{A}_o}$, we construct $\underline{B} = F(\underline{A})$, (for any $\underline{A}$), as follows: Set $B = A^m$, in $\underline{B}$ we interpret any term function f of $\underline{B}_o$ in the following way. By (+) the components f_i ($i < m$) are term functions of $\underline{A}_o$. They can be interpreted in $\underline{A}$ as $(f_i)^{\underline{A}}$. Set $(f^{\underline{B}})_i = (f_i)^{\underline{A}}$. This interpretation preserves the composition of term functions. Hence the identities of $\underline{B}_o$ also hold in $\underline{B}$. Therefore $\underline{B} \in V_{\underline{B}}$. We now extend f to a homomorphism so that it is a functor F. Let $\delta : \underline{A} \longrightarrow \underline{A}'$ be a morphism in the category $\mathfrak{A}$. The image $\varkappa = F(\delta)$ where $\varkappa : \underline{B} \longrightarrow \underline{B}'$, $\underline{B} = F(\underline{A})$ and $\underline{B}' = F(\underline{A}')$ is defined by $\varkappa((a_o,\dots,a_{m-1})) = (\delta(a_o),\dots,\delta(a_{m-1}))$. It is easy to show that F is a functor. The inverse mapping is essentially the projection onto the first coordinate i, however, the domain of an algebra of $\underline{B}$ in $V_{\underline{B}_o}$ is not necessarily a cartesian power, so we have to use its term functions to obtain the effect of the projection. A unary term function of $\underline{B}_o$ is given by $e_o(b_o,b_1,\dots,b_{m-1}) = (b_o,b_o,\dots,b_o)$. Let A be the range of $e_o^{\underline{B}}$. To interpret any term function f of $\underline{A}_i$ in A we first define $\hat{f}$ of $\underline{B}_o$ by $\hat{f}((b_{oo},b_{1o},\dots),(b_{o1},b_{11},\dots),\dots) = (f(b_{oo},b_{1o},\dots),f(b_{oo},b_{1o},\dots),\dots f(b_{oo},b_{1o},\dots))$ and then restrict $\hat{f}^{\underline{B}}$ to A. $\hat{f}^{\underline{B}}$ preserves A because $\hat{f}^{\underline{B}}$ satisfies the identity $e_o\hat{f} = \hat{f}$ in $\underline{B}_o$. We define $\underline{A}$ as the algebra with the carrier A and term functions $\hat{f}^{\underline{B}}/\underline{A}$, where f is a term function of $\underline{A}_o$. Since also $f(g_o,g_1,\dots) = \hat{f}(\hat{g}_o,\hat{g}_1,\dots)$, it follows that Id $\underline{A}_o \subseteq$ Id $\underline{A}$, and hence $\underline{A} \in V_{\underline{A_o}}$. Next we have to show that $G(F(\underline{A})) \cong \underline{A}$ for all $\underline{A} \in V_{\underline{A}}$. The carrier of $\overline{G}(F(\underline{A}))$ is $\underline{\dot{A}} = \{(a,\ldots a,\ldots)/a \in A\}$ and if λ is the mapping from A to $\dot{A}$, then for any term function f of $\underline{A_o}$ it follows: $\lambda(f^{\underline{A}}(a)) = \hat{f}^{\underline{B}}(\lambda(a)),(a \in A^m)$. Thus λ is an isomorphism between $\underline{A}$ and $\underline{\dot{A}}$. It holds $F(G(\underline{B})) \cong \underline{B}$ for any $\underline{B} \in V_{\underline{B}}$. The isomorphism α: $\underline{B} \longrightarrow F(G(\underline{B}))$ is defined for $b \in \underline{B}$ by $\alpha(b) = (e_o(b), e_1(b),\ldots,e_{m-1}(b))$, which is a sequence of the length m. ▲

Let η be an h-elementary relation and let δ be an h-universal relation which is an inverse image of η under a mapping Φ, i.e. for $h \geq 3$, $m \geq 1$, $\Phi: A \longrightarrow E_h{}^m$ $(a_o,\ldots,a_{h-1}) \in \delta$ iff $(\Phi(a_o),\ldots,\Phi(a_{h-1})) \in \eta$. The function $\Phi: A \longrightarrow E_h{}^m$ induces a functor $\hat{\Phi}: V_{\underline{A_\delta}} \longrightarrow V_{\underline{A_\eta}}$ of categories. This is a categorical equivalence. We have only shown that the lattice of congruences of any $\underline{C} \in V_{\underline{A_\delta}}$ is isomorphic to that of $\hat{\Phi}(\underline{C})$; thus $\underline{C}$ is irreducible iff $\hat{\Phi}(\underline{C})$ is so. From this it follows a one-to-one correspondence between the irreducibles of the two varieties.

Without loss of generality we assume, that $E_h{}^m \subseteq A$, and that the function $\Phi: A \longrightarrow E_h{}^m$ is a projection: $\Phi(b) = b$ (for $b \in E_h{}^m$). Now ϕ is a unary term function of $\underline{A_\delta}$. Let $\mathrm{Con}(\underline{C})$ be the congruence lattice of $\underline{C}$.

For any algebra $\underline{C} \in V_{\underline{A_\delta}}$ we define an algebra $\underline{B} = \hat{\Phi}(\underline{C})$ in the following manner: Its carrier is $\overline{B} = \Phi^{\underline{C}}(c) = \{\Phi^{\underline{C}}(c)/c \in \underline{C}\}$. Any n-ary term function f of $\underline{A_\eta}$ is interpreted in $\underline{B}$ by $f(x_1,\ldots,x_n) = f(\Phi(x_1),\ldots,\Phi(x_n))$ and $f^{\underline{B}} = \hat{f}^{\underline{C}}/\underline{B}$. $f^{\underline{B}}$ is a function into $\underline{B}$. An extension to the identities shows that $\underline{B} \in V_{\underline{A_\eta}}$.

<u>Lemma 20 (Knoebel[5]):</u> For any algebra $\underline{C} \in V_{\underline{A_\delta}}$ it holds the isomorphism $\bar{\Phi} = \mathrm{Con}(\underline{C}) \longrightarrow \mathrm{Con}\,\Phi(\underline{C})$, where $\bar{\phi}(\gamma) = \gamma \,\overline{/}\, B$, when $\gamma \in \mathrm{Con}(\underline{C})$, $B = \Phi(\underline{C})$ and $\underline{B} = \hat{\phi}(\underline{C})$.

<u>Corollary 21:</u> An algebra $\underline{C} \in V_{\underline{A_\delta}}$ is subdirectly irreducible if and only if $\hat{\Phi}(\underline{C}) \in V_{\underline{A_\eta}}$ is subdirectly irreducible.

Lemma 22 (Knoebel[5]): If $\underline{B} \in V_{\underline{A}_\eta}$ is a subdirectly irreducible algebra, then there exists a subdirectly irreducible algebra $\underline{C}$ in $V_{\underline{A}_\delta}$ such that $\underline{B} = \hat{\Phi}(\underline{C})$.

Finally we obtain the following theorem:

Theorem 23: Let δ be any relation of B and let $\underline{A}_\delta$ be the algebra generated by δ. In $V_{\underline{A}_\delta}$ there are finitely many subdirectly irreducible algebras, all of them are finite.

Now we are able to establish a common representation theorem for the algebras of varieties generated by a preprimal algebra $\underline{A}$ of one of the forms $\underline{A}_\gamma$, $\underline{A}_\alpha$, $\underline{A}_B$, $\underline{A}_s$, $\underline{A}_\theta$, $\underline{A}_\delta$ (and some algebras $\underline{A}_\rho$). A variety generated by one of these algebras has the following properties:

(i) V has only finitely many subdirectly irreducible elements,

(ii) all subdirectly irreducible elements of V are finite.

Let $\underline{C}_1,\dots,\underline{C}_n$ be all subdirectly irreducible algebras in V, all finite. Then the direct product $\underline{C} = \underline{C}_1 \times \dots \times \underline{C}_n$ is also finite. All subdirectly irreducible algebras $\underline{C}_i$ $(i = 1,\dots,n)$ are homomorphic images of $\underline{C}$ by the projections. By Birkhoff's famous theorem every algebra in a variety is isomorphic to a subdirect product of subdirectly irreducible algebras. It follows that $V_{\underline{A}} = IP_s\{\underline{C}_1,\dots,\underline{C}_n\}$, therefore $V_{\underline{A}} = IP_sH\underline{C}$, i.e. for the varieties $V_{\underline{A}_\gamma}$, $V_{\underline{A}_\alpha}$, $V_{\underline{A}_B}$, $V_{\underline{A}_s}$, $V_{\underline{A}_\theta}$, $V_{\underline{A}_\delta}$ there exists a finite algebra such that $V_{\underline{A}} = IP_sH\underline{C}$.

It arises the question, how many subdirectly irreducible factors will be necessary to construct the finite algebra $\underline{C}$? An answer it is given by the following

Corollary 24: Let V be a variety generated by one of the preprimal algebras $\underline{A}_\gamma$, $\underline{A}_\alpha$, $\underline{A}_B$, $\underline{A}_s$, $\underline{A}_\theta$, $\underline{A}_\delta$. Then there exists a finite algebra $\underline{C}$ which is the direct product of at most two subdirectly irreducible algebras such that $V = IP_sH\underline{C}$.

Proof: $\underline{A}_\gamma$, $\underline{A}_{\{b\}}$ and $\underline{A}_s$ are categorical, i.e. $\underline{C} = \underline{A}_\gamma$, or $\underline{C} = \underline{A}_{\{b\}}$, or $C = \underline{A}_s$. The subdirectly irreducible elements of $\underline{A}_\theta$ are $\underline{A}_\theta$ and $\underline{A}_\theta/\theta$, i.e. $\underline{C} = \underline{A}_\theta$.

$\underline{A_B}$ and $\underline{B}$ (card $B > 1$) are the only subdirectly irreducible algebras of $V_{\underline{A_B}}$ and we have $\underline{C} = \underline{A_B} \times \underline{B}$.

$\underline{A_\alpha}$ and $\underline{A_\alpha^*} = \underline{A^2/\theta}$ (θ uniquely determined by the diagonal), are all subdirectly irreducible algebras in $V_{\underline{A_\alpha}}$, i.e. $C = \underline{A_\alpha^2}$. The subdirectly irreducible algebras in $V_{\underline{A_\delta}}$ (δ a relation from B) are $\underline{A_\delta}$, $\underline{A_0}$ and some homomorphic images of $\underline{A_0}$, i.e. $\underline{C} = \underline{A_\delta} \times \underline{A_0}$. ▲

(The cardinality of $\underline{C}$ is greater or equal to card $A = k$, precisely, there are the following possibilities for card $\underline{C}$: k, $< k^2$, $= k^2$, $\leq k(h l+1)$.)

For a partial order ϱ with a least and a greatest element such that $V_{\underline{A_\varrho}}$ is not congruence distributive, the problem to determine the subdirectly irreducible algebras in $V_{\underline{A_\varrho}}$ distinct from $\underline{A_\varrho}$ is unsolved.

Another problem is the determination of the cardinalities of all finite algebras in a variety, i.e. of the spectrum. In this direction we can answer the following question: For which preprimal algebras the varieties generated by these algebras will have a minimal spectrum ? By 5.2 exactly the preprimal algebras $\underline{A_\alpha}$, $\underline{A_s}$, and $\underline{A_B}$ are paraprimal. Applying Quackenbush's results about varieties with minimal spectrum (see 3.3) we obtain that the varieties $V_{\underline{A_\alpha}}$, $V_{\underline{A_{\{b\}}}}$, $V_{\underline{A_s}}$ are of minimal spectrum. For $V_{\underline{A_s}}$ and $V_{\underline{A_{\{b\}}}}$ it holds $f_V(k^m) = 1$ $(m > 0)$. For $V_{\underline{A_\alpha}}$ it holds $f_V(k^m) = 2$.

5.6 Equationally maximal preprimal algebras

For a preprimal algebra $\underline{A}$ the set of all its term functions is a dual atom in the lattice of all closed classes of functions on A. Let Id $\underline{A}$ be the set of all identities satisfied by an algebra $\underline{A}$. We consider the equational lattice $L_\tau = \{ \text{Id } \underline{A} \mid \underline{A}$ has the type $\tau\}$. For an equationally maximal algebra $\underline{A}$ of the type τ Id $\underline{A}$ is a dual atom in L_τ. Wee see that in the functional lattice preprimal algebras play a similar role as equationally maximal algebras in the equational lattice. Which preprimal algebras are equationally maximal ?

Birkhoff's theorem stating that every algebra is isomorphic to a subdirect product of subdirectly irreducible algebras is useful for studying the equational lattice because

++) If $I \subset K$ and I,K are sets of subdirectly irreducible algebras, then $P_s I \subset P_s K$ and Id $I \supseteq$ Id K.

This is the first technique to determine the equationally maximal preprimal algebras. Preprimal algebras $\underline{A_s}$, $\underline{A_\tau}$, $\underline{A_{\{b\}}}$ are categorical. By 1.4-Theorem 8, these algebras are equationally maximal. Algebras $\underline{A_B}$ (card B >1), $\underline{A_\theta}$, $\underline{A_\omega}$, and $\underline{A_\delta}$($\delta \in B$) are not equationally maximal. This follows from the properties of semi-, hemi-, and paraprimal algebras and from ++).

The second method describes the equationally maximal constantive simple algebras while we know nothing about the subdirectly irreducibles. Which constantive simple preprimal algebra is equationally maximal ? In Foster [3] a condition is given for a constantive simple algebra which is equationally maximal. For this we need the following concept:

Definition 10: A non-trivial finite algebra $\underline{A} = \langle A,F\rangle$ is said to be semiframal if there exist two elements $a,b \in A$ $(a \neq b)$ and a binary term function φ such that $\varphi(a,y) = a$, $\varphi(b,y) = y$ for all $y \in A$.

Which constantive simple preprimal algebras are semiframal ? We have to consider algebras $\underline{A}_\varrho$, $\underline{A}_\tau$, $\underline{A}_\alpha$, and $\underline{A}_\delta$.

We consider the term function f defined by $f(x) = \begin{cases} 0 \text{ if } x=0 \\ 1 \text{ otherwise} \end{cases}$. f is a term function of an algebra $\underline{A}_\varrho$, (ϱ is an arbitrary order relation on A with the least element 0 and the greatest element 1). Since f only can have the values 0 and 1, the function $\min(f(x),y)$ is a term function of $\underline{A}_\varrho$. It follows $\varphi(x,y) = \min(f(x),y) = \begin{cases} 0 \text{ if } x=0 \\ y \text{ if } x \neq 0 \end{cases}$. The term function φ satisfies the identities $\varphi(0,y) = \min(f(0),y) = \min(0,y) = 0$, $\varphi(1,y) = \min(f(1),y) = \min(1,y) = y$, i.e. by Definition 10 algebras of the form $\underline{A}_\varrho$ are semiframal.

The function r defined by $r(x,y) = \begin{cases} c \text{ if } x \text{ is an element of the center of } \tau \\ y \text{ otherwise} \end{cases}$ is a term function of the algebra $\underline{A}_\tau$ since from $(a_0,\dots,c_1,\dots,a_{h-1})$, $(b_0,\dots,b_1,\dots,b_{h-1}) \in \tau$ it follows $(r(a_0,b_0),\dots,r(c_1,b_1),\dots,r(a_{h-1},b_{h-1})) = (b_0,\dots,c_1,\dots,b_{h-1}) \in \tau$ if c_1 is an element of the center and from $(a_0,\dots,a_{h-1}),(b_0,\dots,b_{h-1}) \in \tau$ it follows $(r(a_0,b_0),\dots,r(a_{h-1},b_{h-1})) = (b_0,\dots,b_{h-1}) \in \tau$ if a_i is not an element of the center for all $i=0,\dots,h-1$. The term function r satisfies the identities $r(c,y) = c$ if c is an element of the center and $r(b,y) = y$ if b is a fixed element not belonging to the center. We obtain that algebras of the form $\underline{A}_\tau$ are semiframal.

Algebras of the form $\underline{A}_\alpha$ are not semiframal. If there exist elements a,b, $a \neq b$, $a,b \in A$, and a term function φ in $\underline{A}_\alpha$ with $\varphi(a,y) = a$, $\varphi(b,y) = y$ for all $y \in A$, φ must be quasilinear with respect to an abelian group $\underline{G}$ on A and $\varphi(a+b,b+a) = \varphi(a,b)+\varphi(b,a)-\varphi(0,0) = a+b-\varphi(0,0)$, $\varphi(a+b,a+b) = \varphi(a,a)+\varphi(b,b)-\varphi(0,0) = a+b-\varphi(0,0)$ have to be valid and therefore $a = b$. This is a contradiction.

We consider algebras $\underline{A}_\delta$. Let δ be an h-universal relation on A. Let φ be a binary term function of $\underline{A}_\delta$ such that there exist elements $a,b \in A, a \neq b$ with $\varphi(a,y) = a$, $\varphi(b,y) = y$ for all $y \in A$. There exist elements $a_1,\dots, a_{h-1} \in A$ such that $(a,a_1,\dots,a_{h-1}) \notin \delta$ since, otherwise, δ is a central relation. From $(a,b,\dots,b) \in \delta$ and $(a_1,a_1,a_2,\dots,a_{h-1}) \in \delta$ it follows

$((\varphi(a,a_1), \varphi(b,a_1),\ldots,\varphi(b,a_{h-1})) = (a,a_1,\ldots,a_{h-1}) \notin \delta$. Therefore an algebra $\underline{A}_\delta$ is not semiframal. Concerning semiframal algebras Foster proved:

Lemma 25 (Foster [7]): Let $\underline{A} = \langle A,F\rangle$ be a finite non-trivial constantive and simple algebra. If $\underline{A}$ is semiframal, then $\underline{A}$ is equationally maximal.

Since the algebras $\underline{A}_\gamma$ and $\underline{A}_\varrho$ are constantive, simple, and semiframal we can apply Lemma 25: Every preprimal algebra of the form $\underline{A}_\gamma$ or of the form $\underline{A}_\varrho$ is equationally maximal.

Remark that in [5] Knoebel gave another proof of the equational maximality of $\underline{A}_\varrho$ using D.Scott's ([1]) characterization of an equationally maximal algebra.

Further we obtain the following propositions:

Corollary 26: A constantive simple preprimal algebra is equationally maximal if and only if it is semiframal.

Theorem 27: Let $\underline{A}$ be a preprimal algebra. Then $\underline{A}$ is equationally maximal if and only if one of the following cases is satisfied:

1. $\underline{A}$ is constantive, simple and semiframal,
2. $\underline{A}$ has a one-element subalgebra,
3. $\underline{A}$ has a non-identical automorphism.

5.7 Semidegenerate varieties generated by preprimal algebras

Every variety V contains (up to isomorphism uniquely determined) the one-element or trivial algebra of the type of V. A variety which contains only the one-element algebra is said to be a degenerate variety.

Definition 11 (Lausch/Nöbauer[1]): If no algebra of a variety V consisting of at least two elements has a one-element subalgebra, the variety V is said to be semidegenerate.

Before we consider preprimal algebras generating semidegenerate varieties, we collect some propositions concerning arbitrary algebras which generate semidegenerate varieties. 1.3-Theorem 5 is a Malcev-type-condition for semidegenerate varieties. For instance, let us consider the two-element preprimal algebra $\underline{D_3} = \langle\{0,1\},d,x+y+z,N\rangle$. With n = 1, $p_1(x,y,z) = xy \vee xz \vee yz$, $f_1(x) = x$, $g_1(x) = Nx$ the assumptions of 1.3-Theorem 5 are satisfied and $V_{\underline{D_3}}$ is a semidegenerate variety, i.e. no algebra of $V_{\underline{D_3}}$ has a one-element subalgebra.

The term function φ of a semiframal algebra $\underline{A}$ from Definition 10 satisfies the assumptions of 1.3-Theorem 5, in the following manner: Let $p_1(x,y,z) = \varphi(x,y)$, $p_2(x,y,z) = \varphi(x,z)$, $f_1(x) = g_2(x) = c_b(x) \equiv b$, $f_2(x) = g_1(x) = c_a(x) \equiv a$ where c_a, c_b are term functions of $\underline{A}$ and a,b the elements of A from Definition 10. It holds: $p_1(f_1(x),x,y) = \varphi(b,x) = x$, $p_1(g_1(x),x,y) = \varphi(a,x) = a$, $p_2(f_2(x),x,y) = \varphi(a,y) = a$, $p_2(g_2(x),x,y) = \varphi(b,y) = y$. We obtain:

Corollary 28: If $\underline{A} = \langle A,F\rangle$ is semiframal and the elements $a,b \in A$ with $\varphi(a,y) = a$, $\varphi(b,y) = y$ (Definition 10) are constant term functions of $\underline{A}$, then $\underline{A}$ generates a semidegenerate variety.

For instance, the preprimal algebras $\underline{A_{\varrho}}$ and $\underline{A_{\gamma}}$ generate semidegenerate varieties.

Lemma 29: If $\underline{A}$ is constantive and simple and $V_{\underline{A}}$ semidegenerate, then $\underline{A}$ is equationally maximal.

Proof: Suppose $\underline{A}$ is not equationally maximal and let $\underline{B}$ be a non-trivial algebra and Id $\underline{B} \supset$ Id $\underline{A}$. By 4.2-Lemma 5, the centroid of $\underline{B}$ is a homomorphic image of $\underline{A}$. Since $\underline{A}$ is constantive and simple $C(\underline{B})$ is either isomorphic to $\underline{A}$ or $C(\underline{B}) = \{\beta\}$ is a one-element algebra. Since the centroid of $\underline{B}$ is a minimal subalgebra of $\underline{B}$ and since $\underline{B}$ does not have a one-element subalgebra, $C(\underline{B}) = \{\beta\}$ is impossible. Therefore $C(\underline{B})$ is isomorphic to $\underline{A}$, hence Id $\underline{A}$ = Id $\underline{B}$. From $C(\underline{B}) \subseteq \underline{B}$ it follows Id $C(\underline{B}) \supseteq$ Id $\underline{B}$, i.e. Id $\underline{A}$
Id $\underline{B}$ in contradiction to our assumption. ▲

From Lemma 29 and 1.3-Theorem 5, it follows:

Corollary 30: Let $\underline{A} = \langle A,F\rangle$ be a non-trivial finite algebra and let $\underline{A}$ be simple and constantive. If there exist a natural number n and ternary term functions $p_1,\dots,p_n$ as well as unary term functions $f_1,\dots,f_n,g_1,\dots,g_n$ such that the following identities hold in $V_{\underline{A}}$:

$p_1(f_1(x),x,y) = x,\ p_i(g_i(x),x,y) = p_{i+1}(f_{i+1}(x),x,y),\ (i=1,\dots,n-1),$

then $\underline{A}$ is equationally maximal.

Further it holds:

Lemma 31: If $\underline{A}$ is a non-trivial constantive equationally maximal algebra, then $V_{\underline{A}}$ is semidegenerate.

Proof: Suppose $V_{\underline{A}}$ is not semidegenerate, i.e. there is a non-trivial algebra $\underline{B} \in V_{\underline{A}}$ such that $\underline{B}$ has a one-element subalgebra. Since $\underline{B} \in V_{\underline{A}}$ and $\underline{A}$ is equationally maximal it follows Id $\underline{B}$ = Id $\underline{A}$. By 4.2-Lemma 5, $C(\underline{B})$ is isomorphic to $\underline{A}$. The centroid of $\underline{B}$ is a minimal subalgebra of $\underline{B}$, i.e. $C(\underline{B})$ is the one-element algebra. This is a contradiction. ▲

As next we obtain:

Corollary 32: Let $\underline{A} = \langle A,F\rangle$,($\underline{A}$ non-trivial), be a constantive and equationally maximal algebra. Then every non-trivial algebra $\underline{B} \in V_{\underline{A}}$ has a centroid which is isomorphic to $\underline{A}$, i.e. every algebra $\underline{B} \notin V_{\underline{A}}$ has a minimal subalgebra which is isomorphic to $\underline{A}$.

Preprimal algebras $\underline{A}_{\varrho}$, $\underline{A}_{\tau}$, $\underline{A}_{\alpha}$, $\underline{A}_{\delta}$, ($\delta \in B$), are constantive and simple. There are the following possibilities for these algebras:

1. The variety generated by such an algebra $\underline{A}$ is semidegenerate. Then $\underline{A}$ is equationally maximal and every non-trivial algebra of the variety has the generating algebra $\underline{A}$ as the least subalgebra (centroid).

2. The variety generated by such an algebra is not semidegenerate. Then the algebras are not equationally maximal. In $V_{\underline{A}}$ there are algebras $\underline{B}$ with $C(\underline{B})$ being isomorphic to $\underline{A}$ and algebras with $C(\underline{B})$ being a one-element algebra.

5.8 The lattice of subvarieties of a variety generated by a preprimal algebra

By 5.1-Proposition 1, a preprimal algebra either generates an arithmetical variety or is constantive and simple. Therefore we consider two cases: In the first case, let $\underline{A}$ be a constantive simple preprimal algebra and let $\underline{B} \in V_{\underline{A}}$ ($\underline{B}$ non-trivial). Then Id $\underline{B} \supseteq$ Id $\underline{A}$ and by 4.2-Lemma 5, the centroid $C(\underline{B})$ is a homomorphic image of the centroid $C(\underline{A})$. Since $\underline{A}$ is constantive, $C(\underline{B})$ is a homomorphic image of $\underline{A}$. Since $\underline{A}$ is simple, the centroid of $\underline{B}$ either is isomorphic to $\underline{A}$ or it is the one-element algebra. Let $K \subset V_{\underline{A}}$ be a proper subvariety of $V_{\underline{A}}$ and assume that K contains an algebra $\underline{B}$ which has a centroid isomorphic to $\underline{A}$. Identifying isomorphic algebras we obtain $V_{\underline{A}} \subseteq K$ in contradiction to the assumption. Therefore a proper subvariety of $V_{\underline{A}}$ contains only algebras with a one-element centroid. Now let V_B be the class of all algebras of $V_{\underline{A}}$ which have a one-element centroid. Since every proper subvariety of $V_{\underline{A}}$ contains only algebras with a one-element centroid, any proper subvariety of $V_{\underline{A}}$ is included in V_B. If V_B contains only the one-element algebra, i.e. if $V_B = I$ is the trivial variety, $V_{\underline{A}}$ is a minimal variety and $\underline{A}$ is equationally maximal. The constantive simple preprimal algebras $\underline{A_\varrho}$ and $\underline{A_\gamma}$ are equationally maximal.

In the opposite case V_B is a non-trivial variety ($V_B \neq I$). Let $\underline{B} \in V_B$ and let $\underline{B}' \subseteq \underline{B}$ be a subalgebra of $\underline{B}$. Then $\underline{B}'$ has a one-element centroid since the centroid of $\underline{B}$ is a subalgebra which is a subalgebra of every subalgebra of $\underline{B}$, i.e. $\underline{B}' \in V_B$. If $\underline{B}'$ is a homomorphic image of $\underline{B} \in V_B$, then it holds Id $\underline{B}' \supseteq$ Id $\underline{B}$ and the centroid of $\underline{B}'$ is a homomorphic image of the centroid of $\underline{B}$, i.e. $C(\underline{B}') = \{B\}$ and $\underline{B}' \in V_B$. Let PV_B be the class of all direct products of the algebras of V_B, then Id V_B = Id PV_B, i.e. if $\underline{B}' = P\underline{B}$, then it follows Id $\underline{B}'$ = Id $\underline{B}$ and $C(\underline{B}) \cong C(\underline{B}')$, i.e. $C(\underline{B}')$ is trivial and $\underline{B}' \in V_B$. Thus V_B is a variety. As mentioned above all proper subvarieties of $V_{\underline{A}}$ are subvarieties of V_B.

For the constantive simple preprimal algebras $\underline{A_\varrho}$, $\underline{A_\gamma}$, $\underline{A_\alpha}$, $\underline{A_\delta}$ we obtain the following diagrams where I denotes the trivial variety:

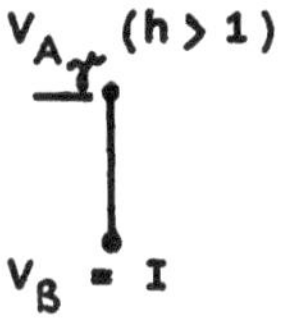

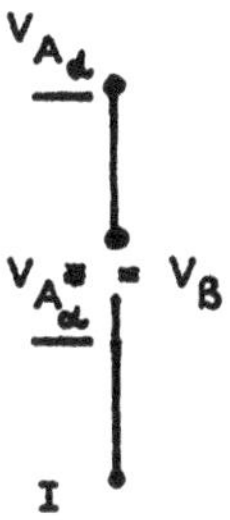

The subdirect irreducible algebras of $V_{\underline{A}_\alpha}$ are $\underline{A}_\alpha$ and $\underline{A}_\alpha^*$. Because $0 = 1 \in \mathrm{Id}\, \underline{A}_\alpha^*$ and $0 = 1 \notin \mathrm{Id}\, \underline{A}_\alpha$, it holds $\mathrm{Id}\, \underline{A}_\alpha \neq \mathrm{Id}\, \underline{A}_\alpha^*$.

We investigate the varieties $V_{\underline{A}_\delta}$ (Knoebel[5]). V_B agrees with the variety $V_{\underline{A}_0}$ generated by the algebra $\underline{A}_0$, where $\underline{A}_0$ corresponds to the algebra $\underline{A_0 \setminus \{B\}} = \langle S_h, S_h \rangle$. The subvarieties of $V_{\underline{A}_0}$ are determined by the proper fully invariant congruences of Id $\underline{A}_0$ and there is a one-to-one correspondence between the proper fully invariant congruences of Id $\underline{A}_0$ and the the normal subgroups of $\underline{S}_h$. (The congruence property in $\underline{A_0 \setminus \{B\}}$ corresponds to the preservation of the left multiplication in $\underline{S}_h$, and the full invariance in $\underline{A_0 \setminus \{B\}}$ corresponds to the preservation of the right multiplication in $\underline{S}_h$.) Therefore in any variety $V_{\underline{A}_\delta}$ there are 5 or 6 subvarieties linearly arranged, each subclass being generated by a single algebra and we obtain for $h \neq 4$

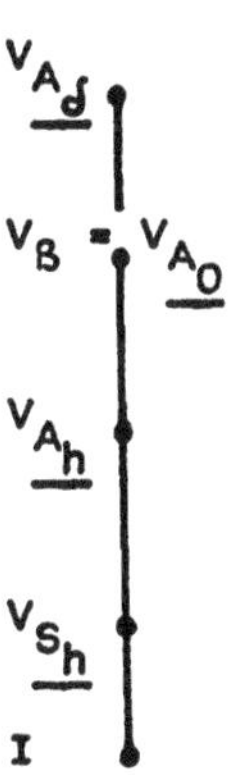

For $h = 4$ between $V_{\underline{A}_4}$ and $V_{\underline{A}_0}$ there is another variety $V_{\underline{V}_4}$. ($\underline{A}_h$ is the group of all odd permutations.)

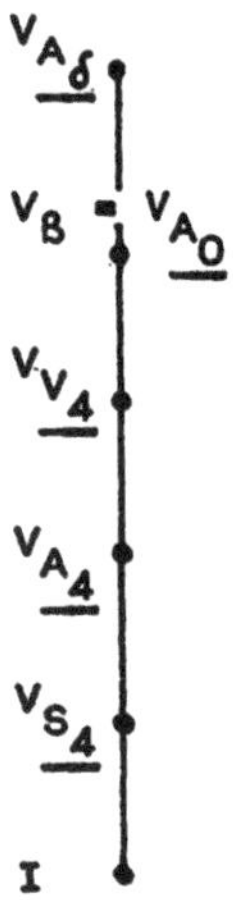

In the other cases, $\underline{A}$ is a preprimal algebra generating an arithmetical variety, i.e. one of the algebras $\underline{A}_s$, $\underline{A}_\theta$, $\underline{A}_B$. $\underline{A}_{\{b\}}$ and $\underline{A}_s$ are equationally maximal. From the properties of semiprimal and linear hemiprimal algebras it follows that $V_{\underline{A}_B}$ (card $B > 1$) and $V_{\underline{A}_0}$ have exactly one non-trivial subvariety, i.e.

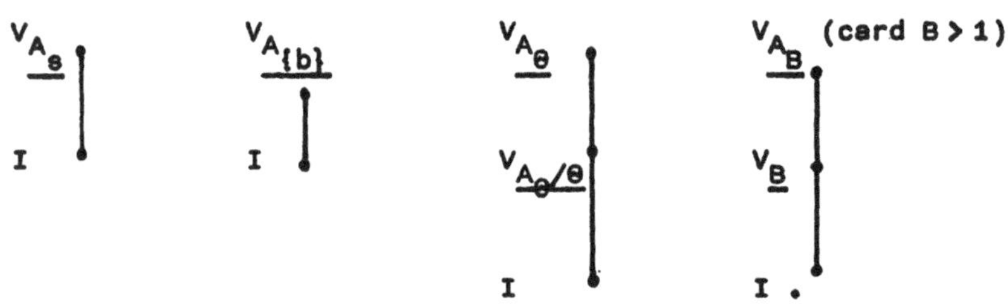

We obtain:

<u>Theorem 33:</u> The lattice of all subvarieties of a variety generated by a preprimal algebra is a chain of the length 2,3,5, or 6.

5.9 On finite axiomatizability of preprimal algebras

By 1.6 a finite algebra which generates a congruence distributive variety is finitely axiomatizable (Baker[2]). Consequently, primal algebras are finitely axiomatizable. Preprimal algebras $\underline{A}_B$, $\underline{A}_s$, and $\underline{A}_\theta$ have finite types (4.4), therefore we have the following

<u>Corollary 34:</u> A preprimal algebra which generates an arithmetical variety is finitely axiomatizable.

Preprimal algebras $\underline{A}_\gamma$ generate congruence distributive varieties and have finite types. Hence algebras $\underline{A}_\gamma$ are finitely axiomatizable.

If ρ is a bounded lattice order on the set A, then by 4.4 $\underline{A}_\rho$ has a finite type. Since $\underline{A}_\rho$ is 2-distributive, it is finitely axiomatizable. For other bounded order relations ρ it is open whether $\underline{A}_\rho$ is finitely axiomatizable or not.

Paraprimal algebras are finitely axiomatizable (1.6). Preprimal algebras $\underline{A}_\alpha$ are paraprimal and have finite types. Therefore algebras $\underline{A}_\alpha$ are finitely axiomatizable.

For algebras $\underline{A}_\delta$ and $\underline{A}_\rho$ (ρ an arbitrary order relation on A with a least and a greatest element) the problem of finite axiomatizability is open. Algebras $\underline{A}_\delta$ have finite types. For an algebra $\underline{A}_\rho$ it is an open problem whether or not $\underline{A}_\rho$ has a finite type.

We remark that D.Pigozzi ([1]) gave an example for a finite constantive equationally maximal algebra which is not finitely axiomatizable.

Problems and comments

1. Find algebraic characterizations for the following classes of preprimal algebras: for all preprimal algebras generating congruence distributive varieties, for all preprimal algebras generating congruence permutable varieties, for all constantive and simple preprimal algebras, for all preprimal algebras.

2. (Knoebel [5]) Give a criterion that separates those order relations with a greatest and a least element which give congruence distributive varieties $V_{\underline{A}_\rho}$ from the remaining ones.

3. Determine all subdirectly irreducible algebras in the variety $V_{\underline{A}_\rho}$.

4. A finite subdirectly irreducible algebra is said to be splitting in a variety if there is a largest subvariety of this variety not containing it. The notion of a splitting algebra (Mc Kenzie [1], Day [1]) arises as a generalization of the lattice-theoretical result of Dedekind and Birkhoff that a lattice is modular respectively distributive if and only if it does not contain the pentagon  respectively the pentagon and the 3-diamond  as a sublattice. $\underline{A}$ is splitting in $V_{\underline{A}}$ if and only if the lattice of all subvarieties of $V_{\underline{A}}$ has exactly one dual atom. Prove that every preprimal algebra $\underline{A}$ is splitting in $V_{\underline{A}}$.

5. Let $\underline{A}$ be a finite non-trivial algebra. The class $QV_{\underline{A}} = ISP\underline{A}$ is said to be a quasivariety generated by $\underline{A}$ (A.I.Malcev [3], Smirnov [1]). Prove that $QV_{\underline{A}_\rho}$, $QV_{\underline{A}_\tau}$ $(h > 1)$, $QV_{\underline{A}_\theta}$, $QV_{\underline{A}_\alpha}$, $QV_{\underline{A}_\delta}$, $QV_{\underline{A}_s}$, $QV_{\underline{A}_{\{b\}}}$ are minimal quasivarieties. Does there exist a finite basis for the equational implications of a quasivariety generated by a preprimal algebra ?

6. A pseudovariety (Eilenberg/Schützenberger [1]) is a class of algebras closed under homomorphic images, subalgebras, and finite direct products. If $\underline{A}$ is finite, then the pseudovariety generated by $\underline{A}$, $HSP_f\underline{A}$, consists of finite algebras only (see also Quackenbush [5]). Characterize pseudovarieties generated by preprimal algebras (Pöschel/Kalužnin [1]).

6. Completeness in universal algebras

Primality and their algebraic and logical consequences are of great interest in the study of truth functions in many-valued logics and in the study of logical circuits in computer science. In this chapter we shall study relations between some concepts of completeness connected with primality. From Rosenberg's theorem (4.1) it follows that in the investigation of the primality preprimal algebras play an essential role.

6.1 Primality

The basic problem in the theory of the superposition of functions on a finite set $A = \{0,\dots,k-1\}$ is the characterization of the subsets $F \subseteq \underline{F}(A)$ for which $\overline{F} = \underline{F}(A)$. Such sets F are called functionally complete in $\underline{F}(A)$. The non-trivial algebras $\underline{A} = \langle A,F\rangle$ where F is functionally complete in $\underline{F}(A)$ are called primal. A special case arises when a functionally complete set F consists of one function only. A function $f \in \underline{F}(A)$ is said to be a Sheffer function if $\{f\}$ is functionally complete in $\underline{F}(A)$ (see also 1.1). We give some examples for primal algebras.

1. (Post[1]) The non-trivial algebra $\underline{A} = \langle A,\min,g\rangle$, $(A = \{0,\dots,k-1\})$, of the type (2,1), where min is the minimum with respect to the ordinary order of the natural numbers and $g(x) = \begin{cases} x+1 & \text{if } x \neq k-1 \\ 0 & \text{if } x = k-1 \end{cases}$, is primal.
2. (Webb[1]) The non-trivial algebra $\underline{A} = \langle A,/\rangle$ of the type (2), where $x/y = \begin{cases} 0 \text{ if } x=y=k-1 \\ \min(x,y)+1 \text{ otherwise} \end{cases}$, is primal.
3. (Foster[1]) The algebra $\underline{A} = \langle A,\cdot,g\rangle$ of the type (2,1) is primal if there exist a binary term function $\cdot$, a unary term function g and an element $0 \in A$ such that

(i) $\langle A\setminus\{0\},\cdot\rangle$ is a group,

(ii) $a \cdot 0 = 0 \cdot a = 0$ for all $a \in A$,

(iii) g is a cyclic permutation on A.

4.(Slupecki[1]) We say that $f \in \underline{F}(A)$ is essential if it depends on at least two variables and takes on all values from A. Thus $f \in \underline{F}(A)$ is not essential if and only if either f depends on at most one variable or some $a \in A$ is not a value of f. An algebra $\underline{A} = \langle A,F\rangle$ (card $A > 2$) is primal if and only if every unary function is a term function of $\underline{A}$ and there exists an essential term function $f \in F$.

The following simple observation is basic in our study:

<u>Lemma 1:</u> A finite non-trivial algebra $\underline{A} = \langle A,F\rangle$ is primal if and only if F is not a subset of all term functions of any preprimal algebra, i.e., of any algebra $\underline{A_\rho}$, $\underline{A_\tau}$ $(h > 1)$, $\underline{A_\theta}$, $\underline{A_\delta}$, $\underline{A_\varepsilon}$, $\underline{A_B}$.

Thus $\underline{A} = \langle A,F\rangle$ is primal if and only if to every maximal class in $\underline{F}(A)$ the set F contains a function f not belonging to that maximal class. This is a very general completeness criterion. The completeness problem is solved by the determination of all maximal classes. Since the number of the maximal classes in $\underline{F}(A)$ is finite and since all maximal classes in $\underline{F}(A)$ are known from the work of Post, Jablonskij and Rosenberg, we have an effective procedure to decise the functional completeness of a given set of functions on a finite set A. O, P, L, C, E, B are the 6 classes of finitary relations occuring in Rosenberg's theorem. If card $A > 2$ from Lemma 1 and 4.1, it follows:

<u>Corollary 2:</u> A finite non-trivial algebra $\underline{A} = \langle A,F\rangle$ is primal if and only if to every relation $\rho \in O \cup P \cup L \cup C \cup E \cup B$ there exists a function $f \in F$ such that f does not preserve ρ

There are several completeness results in the literatur on universal algebra. By (Foster/Pixley[1]) (see also 2.2-Corollary 6 and 1.3-Corollary 4) we obtain:

<u>Corollary 3:</u> An algebra $\underline{A} = \langle A,F\rangle$ (card $A > 1$) is primal if and only if

(i) $\underline{A}$ is simple, has no proper subalgebras and is rigid,

(ii) there is a ternary term function q in $\underline{A}$ such that

$$q(x,x,z) = q(z,x,x) = q(z,x,z) = z.$$

(Using Lemma 1 and our results on preprimal algebras one can give a very simple proof of Corollary 3).

6.2 Primality and preprimality of finite algebras with a single fundamental operation

In case $\underline{A}$ has only one fundamental operation we can formulate 6.1-Corollary 3, in a simpler manner:

Theorem 4 (Rousseau[1]): A finite algebra $\underline{A} = \langle A,f\rangle$ (card $A > 1$) with a single fundamental operation f, where f is at least binary, is primal if and only if $\underline{A}$ is simple, has no proper subalgebras and is rigid.

Proof: Using Rosenberg's classification of all maximal classes in $\underline{F}(A)$ we give Rousseau's proof of Theorem 4. For the 6th class we give a shorter proof than Rousseau's proof. The proof can be restricted to the case that card $A > 2$. By 6.1-Lemma 1, we have to show that f is not a term function of a preprimal algebra if $\underline{A}$ has no proper subalgebras, automorphisms and congruences and conversely, that $\underline{A}$ has no proper subalgebras, automorphisms and congruences if $\underline{A}$ is primal.

If $\underline{A} = \langle A,f\rangle$ is primal, then $f \notin F_\theta$, $f \notin F_s$, $f \notin F_B$, thus $\underline{A}$ has no proper congruences, automorphisms and subalgebras. Conversely, if $\underline{A}$ has no proper subalgebras, automorphisms and congruences then $f \notin F_B$, $f \notin F_s$, $f \notin F_\theta$. Moreover, f cannot be a term function of an algebra $\underline{A}_\varrho$ where ϱ is an order relation on A with the greatest element 1 since from $(a_1,1),\ldots,(a_n,1) \in \varrho$ it follows $(f(a_1,\ldots,a_n),f(1,\ldots,1)) \in \varrho$. If $f(a_1,\ldots,a_n) = 1$ for some $a_1,\ldots,a_n \in A\setminus\{1\}$, we have $f(1,\ldots,1) = 1$, i.e. either $\{1\}$ or $A\setminus\{1\}$ is a proper subalgebra of $\underline{A}$. Suppose card $A = p^m$(p prime number, $m \geq 1$),and $\underline{G} = \langle A,+\rangle$ $(A = \{0,\ldots,k-1\})$ is an abelian group in which each non-zero element has the order p. Since $\underline{G}$ has a basis,$\{b_o,\ldots,b_{m-1}\}$, every element of $\underline{G}$ can uniquely be written in the form $a = a_ob_o+a_1b_1+\ldots+a_{m-1}b_{m-1}$ $(a_o,\ldots,a_{m-1} \in \{0,\ldots,p-1\})$. To every element a it belongs the vector $\mathfrak{a} = (a_o,\ldots,a_{m-1})$. Let $\underline{W}$ be the m-dimensional vector space over the residue class field K_p of the integers modulo p. The mapping given by $a \longrightarrow \mathfrak{a}$ is a group isomorphism of $\underline{G}$ onto $\underline{W}$. Relative to this isomorphism every function $f \in F_n(A)$ corresponds to a function $f(\mathfrak{x}_1,\ldots,\mathfrak{x}_n)$ which is a mapping of the cartesian product $\underline{W}^n$ into $\underline{W}$. A quasilinear function (see 5.2) on $\underline{W}$ (with respect to $\underline{G}$) can be written in the form $f(\mathfrak{x}_1,\dots,\mathfrak{x}_n) = \mathfrak{a} + \mathfrak{x}_1A_1 + \mathfrak{x}_2A_2 + \dots + \mathfrak{x}_nA_n$, where $\mathfrak{a} \in \underline{W}$ and $A_1,\dots,A_n$ are mxm matrices over K_p. If $\det(I-\sum A_i) = 0$, then the equation $\mathfrak{o} = \mathfrak{x}(I-\sum A_i)$ has a non-zero solution $\mathfrak{x} = \mathfrak{x}_o$, so that $f(\mathfrak{x}_o+\mathfrak{x}_1,\dots,\mathfrak{x}_o+\mathfrak{x}_n) = \mathfrak{x}_o + f(\mathfrak{x}_1,\dots,\mathfrak{x}_n)$, i.e. $\mathfrak{x} \longrightarrow \mathfrak{x}_o + \mathfrak{x}$ constitutes a proper automorphism of $\underline{A} = \langle A,f\rangle$. If $\det(I-\sum A_i) \neq 0$, the equation $\mathfrak{a} = \mathfrak{x}(I-\sum A_i)$ has a solution $\mathfrak{x} = \mathfrak{x}_o$ so that $f(\mathfrak{x}_o,\dots,\mathfrak{x}_o) = \mathfrak{x}_o$, i.e. $\{\mathfrak{x}_o\}$ is a proper subalgebra of $\langle A,f\rangle$. Therefore $f \notin F_\alpha$.

Let f be a term function of an algebra $\underline{A}_\gamma$ and let C_o be the centre of the relation γ $(2 \leq h \leq k)$. For each ν let $C_{\nu+1} = C_\nu \cup f(C_\nu^n)$. Then it holds $C_\nu^h \subseteq \gamma$ for each ν . This is clear for $\nu = 0$; suppose $C_\nu^h \subseteq \gamma$ and $(a_o,\dots,a_{h-1}) \in C_{\nu+1}^h$. Then $(a_o,\dots,a_{h-1}) \in \gamma$ if some $a_i \in C_o$. If no $a_i \in C_o$ we may write $a_i = f(a_{1i},\dots,a_{ni})$ with $(a_{1i},\dots,a_{ni}) \in C_\nu^n$ $(i = 0,\dots,h-1)$, so again $(a_o,\dots,a_{h-1}) \in \gamma$ since by induction hypothesis $(a_{1o},\dots,a_{1h-1}),\dots,(a_{no},\dots,a_{nh-1}) \in \gamma$. Since γ is not the unit-relation it follows $\bigcup_{\nu \geq 0} C_\nu \neq A$ and $\bigcup_{\nu \geq 0} C_\nu$ is a proper subalgebra of $\underline{A} = \langle A,f\rangle$.

Let ξ be a primitive h-elementary relation on the set $\{0,\dots,h-1\}$ $(h \geq 3)$, i.e. $(a_o,\dots,a_{h-1}) \in \xi$ if and only if at least two of the a_i are equal. Let F_ξ be the closed class generated by ξ and let $f \in F_\xi$. By 5.5 for f there exist the following possibilities:

(1) f is non-surjective. Assume f has only values of a subset $D \subset \{0,\dots,h-1\}$. Then $\underline{D} = \langle D,f\rangle$ is a proper subalgebra of $\underline{A} = \langle \{0,\dots,h-1\},f\rangle$.

(2) f is a permutation on $\{0,\dots,h-1\}$. In this case f is an automorphism of $\underline{A} = \langle \{0,\dots,h-1\},f\rangle$.

Therefore $f \notin F_\xi$.

Now let η be an h-elementary relation on the set $\{0,\dots,h-1\}^m$ of all m-tuples over $\{0,\dots,h-1\}$ such that $((a_{oo},\dots,a_{om-1}),\dots,(a_{h-1o},\dots,a_{h-1m-1})) \in \eta$ if and only if for each $j < m$ at least two of the elements $a_{oj},a_{1j},\dots,a_{h-1j}$ are equal. η is the m-fold power of the primitive h-elementary relation ξ . Let F_η be the closed class generated by η and $f \in F_\eta$. We consider the algebra $\underline{A} = \langle \{0,\dots,h-1\}^m,f\rangle$ where $f: ((\{0,\dots,h-1\})^m)^n \longrightarrow (\{0,\dots,h-1\})^m$. By 5.5-Lemma 18, $f \in F_\eta$ if and

only if f_i for all $i < m$ is an mn-ary term function of $\underline{A}$. Here $f_i : A^{mn} \longrightarrow A$ is the ith projection of f, i.e. f_i is the composition of e_i and f: $f_i = e_i \circ f$. Since $\underline{A}$ has no proper subalgebra, at least one of the algebras $\underline{B_i} = \langle \{0,\dots,h-1\}, f_i \rangle$ ($i = 0,\dots,m-1$) has no proper subalgebra. Suppose the contrary. Let $\underline{D_o},\dots,\underline{D_{m-1}}$ be proper subalgebras of $\underline{B_o},\dots,\underline{B_{m-1}}$, then the set of all m-tuples $\{(d_o,\dots,d_{m-1}) \mid d_o \in D_o,\dots,d_{m-1} \in D_{m-1}\}$ is a proper subalgebra of $\underline{A}$. Assume $\underline{B_j} = \langle \{0,\dots,h-1\}, f_j \rangle$ has no proper subalgebra. If f_j is non-surjective, we obtain a contradiction. f_j cannot be a permutation on $\{0,\dots,h-1\}$ since f_j is an mn-ary function on $\{0,\dots,h-1\}$ and $m > 1$. Therefore $f_j \notin F_\uparrow$. It follows that $f \notin F_\eta$.
Now let A be a finite set with more than $E_h{}^m$ elements. Let Φ be a function from A onto $E_h{}^m$. Then an h-universal relation on A is a strongly homomorphic preimage of η, i.e. $(a_o,\dots,a_{h-1}) \in \delta$ if and only if $(\Phi(a_o),\dots,\Phi(a_{h-1})) \in \eta$. Let $f \in F_\delta$, then $\Phi(a_o) = \Phi(a_1)$ defines a proper congruence unless $\underline{A}$ has a proper subalgebra. We show that $\Phi(a_{oo}) = \Phi(a_{o1}),\dots,\Phi(a_{n-1o}) = \Phi(a_{n-11})$ imply $\Phi(f(a_{oo},\dots,a_{n-1o})) = \Phi(f(a_{o1},a_{11},\dots,a_{n-11}))$. Suppose the contrary. Since $\Phi(f(a_{oo},\dots,a_{n-1o}))$ and $\Phi(f(a_{o1},a_{11},\dots,a_{n-11}))$ are elements of $\{0,\dots,h-1\}^m$ at least one component of $\Phi(f(a_{oo},\dots,a_{n-1o}))$ is distinct from the corresponding component of $\Phi(f(a_{o1},\dots,a_{n-11}))$. Assume

$e_1(\Phi(f(a_{oo},\dots,a_{n-1o}))) \neq e_1(\Phi(f(a_{o1},\dots,a_{n-11})))$.

If $\underline{A}$ has no proper subalgebra, f is surjective since Φ and e_1 are surjectiv jective. Thus elements $a_{oi},\dots,a_{n-1i}$, $i = 3,\dots,h$ may be choosen so that $e_1(\Phi(f(a_{oi},\dots,a_{n-1i})))$, $(i = 0,\dots,h-1)$, are all distinct, i.e. $(\Phi(f(a_{oo},\dots,a_{n-1o})),\dots,\Phi(f(a_{oh-1},\dots,a_{n-1h})) \notin \eta$.
It follows that

$(f(a_{oo},\dots,a_{n-1o})),\dots,f(a_{on-1},\dots,a_{h-1n-1})) \notin \delta$, i.e. $f \notin F_\delta$. ▲

(Clearly, the condition that $\underline{A}$ is rigid can be restricted to the weaker requirement that $\underline{A}$ is prime rigid, i.e. that $\underline{A}$ has no automorphisms with card A/p cycles of the same prime length p).
Theorem 4 is not only a sharpening of Corollary 3 for algebras with a single fundamental operation, but also says that every maximal class in $\underline{F}(A)$ is a subset of $F_s \cup F_B \cup F_\Theta$. In other words, the maximal classes F_s, F_B, F_Θ form a cover of the set of all maximal classes. By Theorem 4 a

function f not belonging to $F_s \cup F_B \cup F_\theta$ belongs not to another maximal class since $\underline{A} = \langle A,f \rangle$ has no proper subalgebras, automorphisms and congruences and $\underline{A}$ is primal. Thus, in some sense, preprimal algebras which generate arithmetical varieties play a special role. Theorem 4 is a simple algebraic condition for Sheffer functions.

Using Theorem 4, 5.1-Theorem 2 and 5.1-Proposition 1, we receive an algebraic characterization of all preprimal algebras with a single fundamental operation f.

Theorem 5: A finite algebra $\underline{A} = \langle A,f \rangle$ (card $A > 1$) with a single at least binary fundamental operation f is preprimal if and only if $V_{\underline{A}}$ is arithmetical and one of the following cases is satisfied (see also 5.1-Theorem 2):

1. $\underline{A}$ has exactly one subalgebra $\underline{B}$ which is either primal or trivial, $\underline{A}$ is rigid and simple,
2. $\underline{A}$ has no proper subalgebras, the automorphism group of $\underline{A}$ is cyclic and of prime order and $\underline{A}$ is simple,
3. $\underline{A}$ has exactly one non-trivial homomorphic image(this is a primal algebra), $\underline{A}$ has no proper subalgebras and is rigid.

Proof: If $\underline{A} = \langle A,f \rangle$ (card $A > 1$), where f is at least binary, is preprimal, by 5.1-Proposition 1, $\underline{A}$ is either constantive and simple or $\underline{A}$ generates an arithmetical variety. In the first case $\underline{A}$, as a constantive algebra, has no proper subalgebras and no non-identical automorphisms. From Theorem 4 it follows that $\underline{A}$ is primal which contradicts the supposition that $\underline{A}$ is preprimal. Therefore a preprimal algebra with a single fundamental operation generates an arithmetical variety. By 5.1-Theorem 2 one of the cases 1.,2.,3. is satisfied. Conversely, if $\underline{A}$ generates an arithmetical variety and if one of the cases 1.,2. or 3. is satisfied, then $\underline{A}$ is preprimal. ▲

A function f is said to be a pre-Sheffer function if and only if $\overline{\{f\}} \neq \underline{F}(A)$, but for every function $g \in \underline{F}(A) \setminus \overline{\{f\}}$ it holds $\overline{\{f\} \cup \{g\}} = \underline{F}(A)$. Theorem 5 is an algebraic condition for a pre-Sheffer function. The Boolean function f defined by $f(x,y,z) = xNy \vee xNz \vee NyNz$ is an example for a pre-Sheffer function on $\{0,1\}$ since $D_3 = \langle \{0,1\}, f \rangle$ is preprimal (3.1). $f \in \underline{F}(A)$ is said to be a pseudo-Sheffer function if $\underline{A} = \langle A,f \rangle$ is a functionally complete algebra. It is an open problem to characterize pseudo-Sheffer functions.

6.3 Functionally complete algebras

In 1.2 functionally complete algebras were defined as algebras $\underline{A} = \langle A,F\rangle$, $A = \{0,\dots,k-1\}$, $k > 1$, with $\overline{F \cup \{c_i \mid 0 \leq i \leq k-1\}} = \underline{F}(A)$ (c_i constant functions on A). Every function on A is a polynomial function of the algebra $\underline{A}$. From the fact that a preprimal algebra is either quasiprimal or constantive we obtain the proposition:

An algebra $\underline{A} = \langle A,F\rangle$ (card $A > 1$) is primal if and only if

(i) $\underline{A}$ is quasiprimal,

(ii) $\underline{A}$ has no proper subalgebras and is (prime) rigid.

Since a preprimal algebra is quasiprimal if and only if it is functionally complete, we obtain:

<u>Corollary 6:</u> A finite non-trivial algebra $\underline{A} = \langle A,F\rangle$ is primal iff

(i) $\underline{A}$ is functionally complete,

(ii) $\underline{A}$ has no proper subalgebras and is (prime) rigid.

<u>Proof:</u> By Lemma 1 $\underline{A}$ is primal iff F is not contained in one of the maximal classes. By (i) we have $F \not\subseteq F_\varrho$, $F \not\subseteq F_\gamma$ $(2 \leq h \leq \text{card } A)$, $F \not\subseteq F_\alpha$, $F \not\subseteq F_\theta$, $F \not\subseteq F_\delta$ since $\underline{A_\varrho}$, $\underline{A_\gamma}$, $\underline{A_\alpha}$, $\underline{A_\theta}$, $\underline{A_\delta}$ are constantive algebras. By (ii) $F \not\subseteq F_B$, $F \not\subseteq F_s$, therefore $\underline{A}$ is primal. ▲

As a further Corollary by 2.1-Theorem 1 we have:

<u>Corollary 7:</u> $\underline{A} = \langle A,F\rangle$ (card $A > 1$) is primal if and only if

(i) the function q with $q(x,x,z) = q(z,x,x) = q(z,x,z) = z$ is a polynomial function of $\underline{A}$,

(ii) $\underline{A}$ has no proper subalgebras, is simple and rigid.

(This is a weaker primality criterion than Pixley's theorem (Corollary 3)).

Corollary 3 is a condition for algebras which generate arithmetical varieties to be primal. This leads to the question:

Under which conditions is an algebra which generates a congruence permutable variety primal ? If $\underline{A} = \langle A,F\rangle$ generates a congruence permutable variety 5.2 implies $F \not\subseteq F_\varrho$, $F \not\subseteq F_\delta$, $F \not\subseteq F_\gamma$ $(2 \leq h \leq \text{card } A)$. If $\underline{A}$ has no proper subalgebras, is simple and prime rigid, it follows that $F \not\subseteq F_B$,

$F \nsubseteq F_\theta$, $F \nsubseteq F_s$. If $\underline{A}$ is not affine with respect to an elementary abelian p-group, it holds $F \nsubseteq F_\alpha$. Therefore we have:

<u>Corollary 8:</u> $\underline{A} = \langle A,F\rangle$ (card $A > 1$) is primal if and only if

(i) $\underline{A}$ generates a congruence permutable variety,

(ii) $\underline{A}$ has no proper subalgebras, is simple, (prime) rigid and not affine with respect to an elementary abelian p-group.

By a slight modification of Corollary 8 we obtain a necessary and sufficient condition for $\underline{A}$ to be functionally complete.

<u>Corollary 9 (Mc Kenzie[2]):</u> Let $\underline{A} = \langle A,F\rangle$ (card $A > 1$) be an algebra which generates a congruence permutable variety. Then $\underline{A}$ is functionally complete if and only if $\underline{A}$ is simple and not affine with respect to an elementary abelian p-group.

<u>Proof:</u> Since $\underline{A} = \langle A,F\rangle$ generates a congruence permutable variety, $F \nsubseteq F_\xi$, $F \nsubseteq F_\delta$, $F \nsubseteq F_\gamma (h > 1)$. Let $\underline{A}^+ = \langle A,F^+\rangle$ be the algebra obtained from $\underline{A}$ by making each element of $\underline{A}$ the value of a new constant term function. Since $\underline{A}$ and $\underline{A}^+$ have the same congruence lattices, $\underline{A}^+$ generates a congruence permutable variety, hence $F^+ \nsubseteq F_\xi$, $F^+ \nsubseteq F_\delta$, $F^+ \nsubseteq F_\gamma (h > 1)$. Obviously, $\underline{A}^+$ is simple since $\underline{A}$ is simple. Therefore $F^+ \nsubseteq F_\theta$. Since $\underline{A}$ is not affine, it follows that $\underline{A}^+$ is not affine with respect to $\underline{G}$, i.e. $F^+ \nsubseteq F_\alpha$. $\underline{A}^+$ is a constantive algebra, i.e. $\underline{A}^+$ has no proper subalgebras and no non-trivial automorphisms. By Lemma 1 $\underline{A}^+$ is primal, hence $\underline{A}$ is functionally complete.

Let $\underline{A} = \langle A,F\rangle$ be functionally complete, i.e. $\underline{A}^+ = \langle A,F^+\rangle$ primal, hence simple, therefore $\underline{A}$ simple. Suppose $\underline{A}$ is affine with respect to an abelian p-group $\underline{G}$, then $F \subseteq F_\alpha$, i.e. $F^+ \subseteq F_\alpha$ since $\underline{A_\alpha}$ is constantive, therefore by Lemma 1 $\underline{A}^+$ is not primal and $\underline{A}$ not functionally complete.

▲

(We remark that this is another proof of Mc Kenzie's theorem (Mc Kenzie[2]), using the classification of all maximal classes of functions in $\underline{F}(A)$ by Rosenberg. In Mc Kenzie's proof Rosenberg's classification is not needed).

If $\underline{A}$ generates a congruence distributive variety, we have the result, via Fried and Pixley ([1]) that a dual discriminator algebra with at least three elements is functionally complete. It is easy to construct a functionally complete algebra $\underline{A}$ generating a congruence distributive variety, such that $\underline{A}$ is not a dual discriminator algebra.

The following Corollary answers a question of Quackenbush in ([2]).

<u>Corollary 10 (Rosenberg [6]):</u> Let $\underline{A} = \langle A,F\rangle$ (card $A > 2$) be a finite algebra which generates a congruence distributive variety. Then $\underline{A}$ is functionally complete if and only if

(i) $\underline{A}$ is simple,

(ii) $F \nsubseteq F_\tau$ (τ central relation on A, $2 \leq h \leq$ card A),

(iii) $F \nsubseteq F_\rho$ (ρ order relation on A with a greatest and a least element).

<u>Proof:</u> Since $\underline{A} = \langle A,F\rangle$ generates a congruence distributive variety, $F \nsubseteq F_\alpha$, $F \nsubseteq F_\delta$. Let $\underline{A}^+ = \langle A,F^+\rangle$ be the algebra obtained from $\underline{A}$ by making each element of $\underline{A}$ the value of a new constant term function, $\underline{A}^+$ generates a congruence distributive variety, hence $F^+ \nsubseteq F_\alpha$, $F^+ \nsubseteq F_\delta$. $\underline{A}^+$ is simple since $\underline{A}$ is simple, hence $F^+ \nsubseteq F_\Theta$. $\underline{A}^+$ is constantive, i.e. $F^+ \nsubseteq F_B$ $F^+ \nsubseteq F_S$. Since $\underline{A}_\tau$ and $\underline{A}_\rho$ are constantive, it follows $F^+ \nsubseteq F_\tau$, $F^+ \nsubseteq F_\rho$. Therefore, by Lemma 1 it follows that $\underline{A}^+$ is primal and $\underline{A}$ is functionally complete.

Let $\underline{A} = \langle A,F\rangle$ be functionally complete, i.e. $\underline{A}^+ = \langle A,F^+\rangle$ primal, by Lemma 1 it follows that $F^+ \nsubseteq F_\tau$, $F^+ \nsubseteq F_\rho$, $F^+ \nsubseteq F_\Theta$. Since $\underline{A}_\tau$, $\underline{A}_\rho$, $\underline{A}_\Theta$ are constantive, we have $F \nsubseteq F_\tau$, $F \nsubseteq F_\rho$, $F \nsubseteq F_\Theta$. ▲

Finally we obtain a necessary and sufficient condition for any finite non-trivial algebra $\underline{A} = \langle A,F\rangle$ to be functionally complete which is analogous to 6.1-Lemma 1.

<u>Corollary 11:</u> A finite non-trivial algebra $\underline{A} = \langle A,F\rangle$ is functionally complete if and only if

(i) $\underline{A}$ is simple,

(ii) $F \nsubseteq F$, $F \nsubseteq F_\tau (h > 1)$, $F \nsubseteq F_\alpha$, $F \nsubseteq F_\delta$.

<u>Proof:</u> If $\underline{A}$ is functionally complete, $\underline{A}^+$ is primal, i.e. $F^+ \nsubseteq F_\rho$, $F^+ \nsubseteq F_\tau (h > 1)$, $F^+ \nsubseteq F_\delta$, $F^+ \nsubseteq F_\alpha$, $F^+ \nsubseteq F_\Theta$. It follows that $F \nsubseteq F_\rho$, $F \nsubseteq F_\tau (h > 1)$

$F^+ \nsubseteq F_\delta$, $F^+ \nsubseteq F_\alpha$, $F^+ \nsubseteq F_\Theta$. It follows that $F \nsubseteq F_\xi$, $F \nsubseteq F_\gamma (h > 1)$, $F \nsubseteq F_\delta$, $F \nsubseteq F_\alpha$ and $\underline{A}^+$ is simple, thus $\underline{A}$ is simple. If, conversely, $F \nsubseteq F_\xi$, $F \nsubseteq F_\alpha$, $F \nsubseteq F_\gamma$ $(h > 1)$, $F \nsubseteq F_\delta$, it follows that $F^+ \nsubseteq F_\xi$, $F^+ \nsubseteq F_\alpha$, $F^+ \subseteq F_\gamma (h > 1)$, $F \nsubseteq F_\delta$ since $\underline{A}_\xi$, $\underline{A}_\alpha$, $\underline{A}_\gamma$, $\underline{A}_\delta$ are constantive. If $\underline{A}$ is simple, $F \nsubseteq F_\Theta$, therefore $F^+ \nsubseteq F_\Theta$. Since further $F^+ \nsubseteq F_B$ and $F^+ \nsubseteq F_s$, $\underline{A}^+$ is primal, therefore $\underline{A}$ is functionally complete. ▲

We remark that for a functionally complete algebra the classes F_ξ , F_α , $F_\gamma (h>1)$, F_δ ,F_Θ play the same role as F_ξ , F_α , $F_\gamma (h>1)$, F_δ , F_Θ, F_B, F_s for a primal algebra. The algebras $\underline{A_\delta}$, $\underline{A_\xi}$, $\underline{A_\alpha}$, $\underline{A_\gamma} (h > 1)$ $\underline{A_\Theta}$ can be denoted as functionally precomplete algebras. It holds: A finite non-trivial algebra is functionally precomplete if and only if it is a constantive preprimal algebra. From 4.2-Theorem 1 and 4.2-Corollary 2 it follows: Every preprimal algebra is either functionally complete or functionally precomplete. Using Corollary 11 several propositions on functionally complete algebras can be proved. (For functionally precomplete algebras see also (Schweigert [1]).)

Now we will give a condition for a functionally complete algebra $\underline{A} = \langle A,F \rangle$ with $\underline{A}^* = \langle A, F \cup \{c_1\} \rangle$ is primal for any constant function c_1, $c_1 \notin F$, in this case $\underline{A}^*$ is the algebra obtained from $\underline{A}$ by making exactly one element of $\underline{A}$ the value of a new constant term function of $\underline{A}^*$. Such an algebra $\underline{A}$ need not to be preprimal. As example we consider the two-element algebra $\underline{F_4} = \langle \{0,1\} \supset, 1 \rangle$ which is functionally complete since $x \supset 0 = Nx$ and hence $\overline{\{\supset, 0, 1\}} = \underline{F}(\{0,1\})$, but $\underline{F_4}$ is not preprimal.

<u>Theorem 12:</u> Let $\underline{A} = \langle A,F \rangle$ (card $A > 1$) be a functionally complete but non-primal algebra. Then for any c_1 with $c_1(x) \equiv i$, $i \in A$, $c_1 \notin F$, $\underline{A}^* = \langle A, F \cup \{c_1\} \rangle$ is primal if and only if

(i) $\underline{A}$ has at most one subalgebra $\underline{B}$ which is primal or else trivial,

(ii) $\underline{A}$ has no non-trivial automorphisms such that the values of the constant functions $c_1 \notin F$ are invariant elements of the automorphisms.

<u>Proof:</u> Since $\underline{A}$ is functionally complete, $\underline{A}^* = \langle A, F \cup \{c_1\} \rangle$ also is functionally complete. Suppose $\underline{A}$ has a trivial or primal subalgebra $\underline{B}$. It follows that Id $\underline{B} \supseteq$ Id $\underline{A}$ and thus (by 4.2-Lemma 5) the centroid of $\underline{B}$ is

a homomorphic image of the centroid of $\underline{A}$. Since $\underline{B}$ is primal or trivial, $C(\underline{B}) = \underline{B}$. Therefore the centroid of $\underline{A}$ is not empty. Since the centroid of $\underline{A}$ is the least subalgebra of $\underline{A}$ and since $\underline{A}$ has at most one proper subalgebra, it holds $C(\underline{A}) = \underline{B}$. It follows that the value of c_i in $\underline{A} = \langle A, F \cup \{c_i\} \rangle$ is not in B and hence $\underline{A}^*$ has no proper subalgebra. Suppose $\underline{A} = \langle A, F \cup \{c_i\} \rangle$ has a non-trivial automorphism s, s also is an automorphism of $\underline{A}$ and it holds $c_i(s(x)) = s(c_i(x))$, i.e. $s(i) = i$ and i is an invariant element of s, therefore s is not an automorphism of $\underline{A}$ and $\underline{A} = \langle A, F \cup \{c_i\} \rangle$ has no non-trivial automorphisms.

$\underline{A}^*$ is functionally complete, has no proper subalgebras, and is rigid. By Corollary 6, $\underline{A}^*$ is primal.

Let $\underline{A}^* = \langle A, F \cup \{c_i\} \rangle$ be a primal algebra for any $c_i \notin F$. Let $\underline{B}$ be a non-trivial subalgebra of $\underline{A}$ and let $c_i \notin F$ $(c_i(x) \equiv i)$. It follows that $i \notin B$ for otherwise, $\underline{A}^* = \langle A, F \cup \{c_i\} \rangle$ would have the subalgebra $\underline{B}$. Only the values of the constant term functions of $\underline{A}$ are elements of the subalgebra, i.e. $\underline{B}$ is the centroid of $\underline{A}$ if there exists a non-empty centroid. Since $\underline{A}$ is functionally complete $\underline{B}$ is also functionally complete since the ternary discriminator t of A (2.1) is also the discriminator of B and a polynomial function of $\underline{B}$. Being a constantive algebra $\underline{B}$ has no proper subalgebra and is rigid. Therefore $\underline{B}$ is primal.

Let s be a non-trivial automorphism of $\underline{A}$ and let $c_j \in F$. It holds $s(c_j(x)) = c_j(s(x))$, i.e. $s(j) = j$, i.e. all elements of the centroid of $\underline{A}$ are invariant elements of s. Let $c_i \notin F$, then it holds $s(c_i(x)) \neq c_i(s(x))$, i.e. $s(i) \neq i$ since $\underline{A}^+ = \langle A, F \cup \{c_i\} \rangle$ is primal. ▲

Examples for algebras $\underline{A}$ covered by Theorem 12 are the functionally complete preprimal algebras. The conditions (i) and (ii) are satisfied since functionally complete preprimal algebras are of the form $\underline{A_s}$ or $\underline{A_B}$.

6.4 A new proof of Rosenberg's primal algebra characterization theorem by R.W.Quackenbush

In 6.3 algebraic characterization theorems for primal or functionally complete algebras were obtained as simple corollaries of Rosenberg's primal algebra characterization theorem and of the properties of preprimal algebras. It arises the question, whether one can obtain a simpler or at least a better understandable structural proof of Rosenberg's theorem using universal algebraic theorems by Foster, Pixley, Mc Kenzie, Quackenbush and others. The difficult part of Rosenberg's original proof is the assertion that all preprimal algebras have been found there. This is equivalent to the proposition: If for an algebra $\underline{A} = \langle A,F \rangle$ (card $A > 2$) F is not included in one of the classes F_{ξ}, F_{α}, $F_{\gamma}(h > 1)$, F_B, F_{θ}, F_s, F_{δ}, then $\underline{A} = \langle A,F \rangle$ is primal or equivalent to the proposition: If no relation in one of the 6 classes O,P,L,C,E,B (4.1) is a subalgebra of a power of $\underline{A}$, then $\underline{A}$ is primal.

The basic idea of the proof in (Quackenbush[4]) is the following: By 2.2-Corollary 6, $\underline{A} = \langle A,F \rangle$ is primal if and only if $\underline{A}$ is simple, rigid, has no proper subalgebras and generates a congruence permutable and congruence distributive (= arithmetical) variety. The first three properties hold for $\underline{A}$ since $F \nsubseteq F_B$, $F \nsubseteq F_s$, $F \nsubseteq F_{\theta}$. If we also know that $\underline{A}$ generates a congruence permutable variety and is not affine in relation to an elementary abelian p-group from Mc Kenzie's theorem (6.3-Corollary 8), it follows that $\underline{A}$ is primal. Since $F \nsubseteq F_{\alpha}$, the second condition is satisfied. (The proof of 6.3-Corollary 8 in Mc Kenzie[2]) is independent of Rosenberg's theorem).

We have to prove that $\underline{A} = \langle A,F \rangle$ generates a congruence permutable variety if F is not included in one of the maximal classes and we make use of the following theorem (Quackenbush[3]): If every subalgebra of a finite power of $\underline{A}$ (card $A = k$) has a power of k as its cardinality, then $\underline{A}$ generates a congruence permutable variety. The proof of Rosenberg's theorem follows from

Theorem 13: Let $\underline{A} = \langle A,F \rangle$ ($A = \{0,\ldots,k-1\}$), $k > 2$, be an algebra such that no subalgebra of a finite power of $\underline{A}$ belongs to one of the classes O,P,L,C,E,B. Then every subalgebra of a finite power of $\underline{A}$ has a power of k as its cardinality.

Quackenbush's proof of Theorem 13 is included in the Appendix.

Problems and comments

1. Give an algebraic characterization for a pseudo-Sheffer function (see 6.2).

2. Give proofs of the following well-known statements. Apply 6.3-Corollary 11.
a. A finite non-trivial algebra $\underline{A}$ is functionally complete if and only if the ternary discriminator function of A is a polynomial function of $\underline{A}$.
b. There is no functionally complete lattice or semilattice.

3. A finite non-trivial algebra $\underline{A} = \langle A,F \rangle$ is said to be n-primal for some n if every n-ary function $f: A^n \longrightarrow A$ is a term function of $\underline{A}$. Therefore $\underline{A}$ is primal iff $\underline{A}$ is n-primal for any positive integers n. As well-known for an arbitrary algebra $\underline{A}$ only the following three cases are possible (see also 4.2): a. $\underline{A}$ is 2-primal and thus n-primal for all positive integers n, i.e. $\underline{A}$ is primal, b. $\underline{A}$ is 1-primal, but not n-primal for any $n > 1$, c. $\underline{A}$ is n-primal for no positive integer n. The term functions of all 1-primal algebras were determined in (Burle [1]) and in (Rosenberg [4]). Every 1-primal algebra is simple, has no proper subalgebras and no non-trivial automorphisms.

7. Examples for preprimal algebras

In 1966 Y.Imai and K.Iseki ([1]) defined a class of algebras of the type (2,0), the so-called BCK-algebras. Since some finite subdirectly irreducible elements of the variety of bounded commutative BCK-algebras are preprimal algebras, we can apply our results to these algebras.

The notion of a BCK-algebra originates from two different ways. One of the motivations is based on set theory. Another motivation comes from classical and non-classical propositional calculi. There are some systems of propositional calculi which contain only the implication functor among the logical functors. These examples are the systems of the classical implicative calculus, the positive implicative calculus, the weak positive implicative calculus and others (see also (Rasiowa[1]).

An algebra $\underline{X} = \langle X, *, 0\rangle$ of type (2,0) is said to be a BCK-algebra provided that the following axioms are satisfied:

(i) $((x * y) * (x * z)) * (z * y)) = 0$,

(ii) $(x * (x * y)) * y = 0$,

(iii) $x * x = 0$,

(iv) $0 * x = 0$,

(v) If $x * y = 0$ and $y * x = 0$, then $y = x$.

The class of all BCK-algebras is a quasi-variety (A.I.Malcev[3]). The subclasses of the commutative, of the positive implicative, and of the implicative BCK-algebras are varieties (Iseki/Tanaka[1]). A BCK-algebra is said to be an implicative BCK-algebra if the axioms (i),(ii), (iii),(iv),(v), and

(vi) $x * (y * x) = x$ are satisfied.

are satisfied. An axiom system for the variety of the implicative BCK-algebras is given by

(1) $x * x = 0$,

(2) $x * 0 = x$,

(3) $y * (y * x) = x * (x * y)$,

(4) $(z * y) * x = (z * x) * (y * x)$.

This variety is generated by the two-element algebra $\underline{F_8} = \langle \{0,1\} * ,0\rangle$ which is dual isomorphic to the algebra $\underline{F_4}$ in 3.2 (Denecke [3]). The class of the implicative BCK-algebras describes the classical implicative propositional calculus in the sense of Rasiowa [1].

An algebra $\underline{X} = \langle X, *, 0,1\rangle$ $(0 \neq 1)$ of type (2,0,0) is said to be a bounded commutative BCK-algebra provided that the following axioms are satisfied (Traczyk [2]):

(a) $(x * y) * z = (x * z) * y,$

(b) $x * (x * y) = y * (y * x),$

(c) $x * x = 0,$

(d) $x * 0 = 0,$

(e) $x * 1 = 0.$

Examples for term functions of any bounded commutative BCK-algebra are $\wedge$ and $\vee$ defined by $x \wedge y = x * (x * y)$ and $x \vee y = 1 * ((1 * x) * (1*y))$. Traczyk ([2]) proved that every bounded commutative BCK-algebra is a distributive lattice with respect to the order defined by $x \leq y$ if and only if $x * y = 0$.

The greatest lower bound denoted by $\wedge$ satisfies the equation $x \wedge y = x * (x * y)$ and the least upper bound denoted by $\vee$ satisfies the equation $x \vee y = 1 * ((1 * x) * (1 * y))$. From 1.3-Corollary 3 and $d(x,y,z) = (x\wedge y) \vee ((x\wedge z) \vee (y\wedge z))$ it follows that the variety of the bounded commutative BCK-algebras is congruence distributive. Traczyk ([2]) proved that every subdirectly irreducible bounded commutative BCK-algebra is a chain, i.e. for any elements a,b of the algebra it holds either $a * b = 0$ or $b * a = 0$. Especially in this paper all finite subdirectly irreducible commutative BCK-algebras were characterized.

Theorem 1: For every natural number $n > 1$ there exists a subdirectly irreducible bounded commutative BCK-algebra with exactly n elements. It is unique up to isomorphisms and represented by the algebra $\underline{X_n} = \langle \{0,\ldots,n-1\} * ,0,n-1\rangle$ with $m * k = \begin{cases} m-k & \text{if } k < m \\ 0 & \text{otherwise} \end{cases}$.

The order relation defined by $x \leq y$ if and only if $x * y = 0$ for any bounded commutative BCK-algebra is the ordinary order of the natural numbers. From Theorem 1 and the congruence distributivity of the variety of bounded commutative BCK-algebras it follows that this variety is not generated by a finite algebra. Otherwise we should have a finite algebra $\underline{A}$ with $V_{\underline{A}} = IP_s HS\underline{A}$. Since this representation is true for all algebras $\underline{X_n}$ from Theorem 1 and since for every integer $n > 1$ there exists a subdirectly irreducible algebra $\underline{X_n}$, this is impossible. Next we state:

<u>Theorem 2:</u> Every finite subdirectly irreducible bounded commutative BCK-algebra is semiprimal.

<u>Proof:</u> Using 2.2-Theorem 3, and 2.1-Theorem 1, we verify that every algebra $\underline{X_n}$ from Theorem 1 is quasiprimal, i.e. that the ternary discriminator is a term function of every algebra $\underline{X_n}$, that the subalgebras of $\underline{X_n}$ have no proper automorphisms, and that no two distinct subalgebras of $\underline{X_n}$ with more than one element are isomorphic. At first we construct term functions of $\underline{X_n}$ by which the ternary discriminator t can be expressed by superposition. We apply results of Rosser/Turquette ([1]) or Denecke ([6]). In these papers it is shown that the functions j_r with $j_r(x) = \begin{cases} 0 \text{ if } x=r \\ n-1 \text{ if } x \neq r \end{cases}$ and $0 \leq r \leq n-1$ are term functions of $\underline{X_n}$. It follows that

$t(x,y,z) = (x \wedge j_o((x \vee y) * (x \wedge y))) \vee (z \wedge ((n-1) * j_o((x \vee y) * (x \wedge y))))$.

Indeed for $x \neq y$ we have $x \wedge y < x \vee y$; therefore $j_o((x \vee y) * (x \wedge y)) = n-1$,

$t(x,y,z) = (x \wedge (n-1)) \vee (z \wedge ((n-1) * (n-1))) = x \vee (z \wedge 0) = x \vee 0 = x$

and for $x=y$ we have $x \wedge y = x \vee y$; therefore $j_o((x \vee y) * (x \wedge y)) = 0$ and

$t(x,x,z) = (x \wedge 0) \vee (z \wedge ((n-1) * 0)) = o \vee (z \wedge (n-1)) = 0 \vee z = z$.

The next step is to prove that the subalgebras of $\underline{X_n}$ have no proper automorphisms and no two distinct subalgebras with more than one element are isomorphic. The following functions are term functions of $\underline{X_n}$:

p_a defined by $p_a(x) = \begin{cases} a & \text{if } x=a \\ n-1 & \text{otherwise} \end{cases}$ $(0 \leq a \leq n-1)$ since $p_a(x) = x \vee j_a(x)$,

f_a defined by $f_a(x,y) = \begin{cases} a \text{ if } y=a \\ x \text{ if } y \neq a \end{cases}$ since $f_a(x,y) = p_a(y) * (((n-1) * x) \wedge j_a(y))$.

Suppose φ is a non-identical automorphism of a subalgebra $\underline{B} \subseteq \underline{X_n}$. Then for some $a \in B$ it holds $\varphi(a) = b \neq a$. Let $\varphi(b) = b'$. Then $b \neq b'$ since φ

is an automorphism of $\underline{B}$. There exists a term function f_b of $\underline{B}$ defined by $f_b(x,y) = \begin{cases} b & \text{if } y=b \\ x & \text{if } y \neq b \end{cases}$ for all $x,y \in B$. Since φ is an automorphism, we have $f_b(\varphi(a), \varphi(b)) = \varphi(a) = \varphi(f_b(a,b)) = \varphi(b)$. But this is false in view of $a \neq b$, $b \neq b'$ and the definition of f_b.

We verify that no two distinct subalgebras of more than one element are isomorphic. Let $\underline{A_1}$ and $\underline{A_2}$ be two distinct subalgebras of $\underline{X_n}$ and suppose $\underline{A_1}$ is isomorphic to $\underline{A_2}$ under the isomorphism φ. Since $\underline{X_n}$ is finite, no proper subalgebra of $\underline{A_2}$ can be isomorphic to $\underline{A_2}$ and hence we may suppose the set $A_1 \setminus A_2$ is non-empty. Let $a \in A_1 \setminus A_2$ and suppose that, under the isomorphism φ we have $\varphi(a) = b \in A_2$. Further let $c \neq a$ be any other element of A_1 and suppose $\varphi(c) = d \in A_2$, $d \neq b$. We consider the above defined term function f_a of $\underline{X_n}$. We have $\varphi(f_a(c,a)) = f_a(\varphi(c), \varphi(a))$. It holds $\varphi(f_a(c,a)) = b$, but $f_a(\varphi(c), \varphi(a)) = f_a(d,b) = d$ and $d \neq b$. Therefore $\varphi(f_a(c,a)) \neq f_a(\varphi(c), \varphi(a))$ is only possible if $a = b$, but $a = b$ is impossible since $a \in A_1 \setminus A_2$, $b \in A_2$; therefore, $f(d,b) = d = b$ which is a contradiction. ▲

Let V_{FSI} denote the class of all finite subdirectly irreducible members of the variety of bounded commutative BCK-algebras. Let us identify every two isomorphic algebras of V_{FSI}. By Theorem 1, V_{FSI} is an enumerable set. For every $\underline{X_m}$, $\underline{X_n} \in V_{FSI}$ define

$\underline{X_m} \leq \underline{X_n}$ if and only if $\underline{X_m}$ is a subalgebra of $\underline{X_n}$.

By this definition $\langle V_{FSI}, \leq \rangle$ is an ordered set and by (Traczyk[2]) it holds:

<u>Corollary 3:</u> $\langle V_{FSI}, \leq \rangle$ is a distributive lattice, isomorphic to the lattice $\langle N, / \rangle$ of all natural numbers, where $m-1/n-1$ means that $m-1$ divides $n-1$.

It holds $\underline{X_m} \subseteq \underline{X_n}$ if and only if $m-1/n-1$. From $m-1/n-1$ it follows that there exists a natural number $r > 0$ with $n-1 = r(m-1)$.

$\langle \{0, r, \dots, (m-1)r\}, *, 0, n-1 \rangle$ is a subalgebra of $\underline{X_n}$ since

$\{0, r, \dots, (m-1)r\} \subseteq \{0, 1, \dots, n-1\}$ and

$n_1r * n_2r = \begin{cases}(n_1-n_2)r & \text{if } n_1 > n_2 \\ 0 & \text{otherwise}\end{cases}$ $,(n_1-n_2)r, 0 \in \{0,r,\dots,(m-1)r\}$.
$\langle\{0,r,\dots,(m-1)r\},*,0,n-1\rangle$ is isomorphic to $\underline{X}_m$. Suppose that $\underline{X}_n$ has a subalgebra isomorphic to $\underline{X}_m$:

$$\langle\{e_0,\dots,e_{m-1}\},*,e_0,e_{m-1}\rangle \subseteq \langle\{0,\dots,n-1\},*,0,n-1\rangle$$

with $e_0 = 0$, $e_{m-1} = n-1$ and $e_0 < e_1 < \dots < e_{m-2} < e_{m-1}$. We shall prove by induction that $e_i = ie_1$ for all $0 \leq i \leq m-1$. If $i = 2$ one can use $e_2 * e_1 = e_2-e_1 = e_1$ since $e_0 < e_2-e_1 < e_2$. Assume the proposition to be valid for every i such that $0 \leq i < m-1$ and prove it for $i = m-1$. We have $e_{m-1} * e_{m-2} = e_{m-1}-e_{m-2} < e_{m-1}$. By our assumption of induction $e_{m-1}-e_{m-2} = e_r, 0 < r < m-1$. It follows that $e_{m-1} = (m-2)e_1 + re_1 = (m-2+r)e_1$. Suppose $r > 1$. Then we should have $e_{m-1} * e_{r-1} = e_{m-1}-e_{r-1} = (m-2+r)e_1-(r-1)e_1 = (m-1)e_1$, i.e. $e_{m-2} < e_{m-1} * e_{r-1} < e_{m-1}$. This is a contradiction and one has $r = 1$ and $e_{m-1} = (m-1)e_1$. From $e_{m-1} = n-1$ we obtain $n-1 = (m-1)e_1$ and $m-1/n-1$.

As a consequence of Corollary 3 every finite subdirectly irreducible bounded commutative BCK-algebra has a subalgebra isomorphic to $\underline{X}_2 = \langle\{0,1\},*,0,1\rangle$. $\underline{X}_2$ is a primal algebra. Since a semiprimal algebra is preprimal if and only if it has exactly one subalgebra which is either trivial or primal and since every bounded commutative BCK-algebra has $\underline{X}_2$ as a subalgebra, exactly those algebras $\underline{X}_n$ are preprimal which have only $\underline{X}_2$ as a subalgebra.

<u>Corollary 4:</u> The algebra $\underline{X}_n$ is preprimal if and only if $n-1 = p$ is a prime number.

The term functions of $\underline{X}_{p+1}$ are all functions defined on $\{0,\dots,p\}$ which preserve the subset $\{0,p\}$. The variety $V_{\underline{X}_{p+1}}$ has exactly one subvariety, namely $V_{\underline{X}_2}$.

Appendix

Proof of 6.4-Theorem 13 (Quackenbush [4])

Let $\underline{A} = \langle A,F\rangle$ be a finite non-trivial algebra ($A = \{0,\dots,k-1\}$) such that no subalgebra of a finite power of $\underline{A}$ belongs to one of the classes O,P,L,C,E,B of Rosenberg's classification. Suppose that for some natural number m $\underline{B}$ is a subalgebra of $\underline{A}^m$ such that card B is not a power of card A = k, and suppose that for all $n < m$ every subalgebra of $\underline{A}^n$ has a power of k as its cardinality. We shall construct a contradiction to this assumption.

It holds $m > 1$ since a proper subalgebra of $\underline{A}$ would be a unary central relation. The first step is to prove that $\underline{B}$ is totally reflexive and totally symmetric if $m \geq 4$ (see 4.1). Then except of $\underline{B}$ being totally reflexive and totally symmetric the cases m = 2 and m = 3 as possibilities are eliminated. In the last section of the proof the totally reflexive and totally symmetric case is considered.

I. The first step of the proof

Let $E = \{i_1,\dots,i_j\}$ be a subset of $\{1,\dots,m\}$, $i_1 < \dots < i_j$. We define $\pi_E\colon A^m \longrightarrow A^j$ by $\pi_E(a_1,\dots,a_m) = (a_{i_1},\dots,a_{i_j})$. For $1 \leq i \leq m$, let $p_i = \pi_{E(i)}$ where $E(i) = \{1,\dots,i-1,i+1,\dots,m\}$.

Proposition 1.1: For $1 \leq i \leq m$ it holds $\operatorname{card} p_i(b) = k^{m-1}$.

Proof: By induction over card E we shall show:

(+) $\operatorname{card} \pi_E(B) = k^{\operatorname{card} E}$.

Since $\underline{A}$ has no proper subalgebras, it follows $\pi_{\{i\}}(B) = A$ for $1 \leq i \leq m$. Let $n < m$, and suppose that (+) holds for all subsets E of $\{1,\dots,m\}$ with card $E < n$. Now let E be an n-element subset of $\{1,\dots,m\}$. Since $\pi_E(B) \subseteq A^n$ and $n < m$, it holds $\operatorname{card} \pi_E(B) = k^i$ for some i. Obviously, $\operatorname{card} \pi_E(B) \leq k^n$, and by our induction hypothesis, $\operatorname{card} \pi_E(B) \geq k^{n-1}$. Hence, $\operatorname{card} \pi_E(B) = k^{n-1}$ or k^n. Let $i \in E$ and $E' = E \setminus \{i\}$. Then $\operatorname{card} \pi_E(B) = k^{n-1}$ implies that $p_i(\pi_E(B)) = \pi_{E'}(B)$, i.e. if $b_j = b_j'$, then $b_i = b_i'$ for $b,b' \in B$ (by b_i we denote the i th coordinate of b). Thus for any

$b, b' \in B$, $p_i(b) = p_i(b')$ implies that $b = b'$; thus p_i is an embedding of $\underline{B}$ into $\underline{A}^{m-1}$. This contradicts our assumption that every subalgebra of $\underline{A}^{m-1}$ has a power of k as its cardinality. Thus card $\pi_E(B) = k^n$. ▲

<u>Corollary 1.2 :</u> $k^{m-1} <$ card $B < k^m$

For a partition P of $\{1,\dots,m\}$ let $B_P = \{ b \in B \mid$ if i and j belong to the same P-class, then $b_i = b_j\}$, $\langle B_P, F\rangle$ is a subalgebra of $\underline{B}$. We shall denote a partition by listing only its non-trivial classes; e.g. (i,j) denotes the partition with exactly one non-trivial class {i,j} .

(++) Notice that $\langle B_P, F\rangle$ can be embedded into A^{m-1} for non-trivial P, therefore card B_P is a power of k.

<u>Proposition 1.3:</u> Let $m \geq 4$. If card $B_{(i,j)} = k^{m-1}$ for some $1 \leq i < j \leq m$, then card $B_{(i,j)} = k^{m-1}$ for all $1 \leq i < j \leq m$.

<u>Proof:</u> Without loss of generality let card $\underline{B}_{(1,2)} = k^{m-1}$; it suffices to prove that card $B_{(2,3)} = k^{m-1}$. Since card $B_{(1,2)} = k^{m-1}$, we have card $B_{(1,2,3)} = k^{m-2}$. By Proposition 1.1 and (++) we have card $B_{(2,3)} \in \{k^{m-1}, k^{m-2}\}$. Suppose that card $B_{(2,3)} = k^{m-2}$. Then $B_{(1,2,3)} = B_{(2,3)}$, i.e. for $b \in B$, $b_2 = b_3$ implies that $b_1 = b_2$. But card $\pi_{\{1,2,3\}}(B) = k^3$ by Proposition 1.1, hence, for $a_1 \neq a_2$ there exists $b \in B$ such that $b_1 = a_1$, $b_2 = b_3 = a_2$. Therefore, $B_{(2,3)} \neq B_{(1,2,3)}$, and so card $B_{(2,3)} = k^{m-1}$. ▲

Define $B' \subseteq A^m$ by

$B' = \{ (a_2, a_3, \dots, a_m, a_m') \mid (a_1, \dots, a_{m-1}, a_m), (a_1, \dots, a_{m-1}, a_m') \in B$ for some $a_1 \in A \}$.

Then $\langle B', F\rangle$ is a subalgebra of $\underline{A}^m$.

<u>Proposition 1.4:</u> Let $m \geq 4$. Suppose card $B_{(2,3)} = k^{m-2}$. Then card $B'_{(m-1,m)} = k^{m-1}$; $k^{m-1} <$ card $B' < k^m$, and card $B'_{(1,2)} = k^{m-2}$.

<u>Proof:</u> By Proposition 1.1 we have card $p_1(B) = k^{m-1}$; hence, card $B'_{(m-1,m)} = k^{m-1}$. Now, $B'_{(m-1,m)} \subseteq B'$ implies card $B' \geq k^{m-1}$. Since card $p_m(B) = k^{m-1}$, card $B > k^{m-1}$ implies that there are $a_1, \dots, a_{m-1}, a_m, a_m' \in A$ such that $a_m \neq a_m'$ and $(a_1, \dots, a_{m-1}, a_m), (a_1, \dots, a_{m-1}, a_m') \in B$. Hence $(a_2, \dots, a_m, a_m'), (a_2, \dots, a_m', a_m) \in B'$ and $a_m \neq a_m'$, consequently, card $B' > k^{m-1}$. Since $p_m(B) = k^{m-1}$, for any $a_1, \dots, a_{m-1} \in A$ there is an element $a_m \in A$ such that $(a_1, \dots, a_{m-1}, a_m) \in B$. The cardinality restriction

card $B_{(2,3)} = k^{m-2}$ implies that a_m is unique if $a_2 = a_3$. Thus $(a_3,a_3,a_4,\dots,a_{m-1},a_m,a_m') \notin B'$ for $a_m \neq a_m'$. Hence, card $B' < k^m$, and since card $B_{(2,3)} = k^{m-2}$, we have card $B'_{(1,2)} = k^{m-2}$. ▲

<u>Proposition 1.5:</u> Let $m \geq 4$ and $1 \leq i < j \leq m$; then card $B_{(i,j)} = k^{m-1}$.

<u>Proof:</u> If card $B_{(2,3)} = k^{m-1}$, then is nothing to prove by Proposition 1.3, Suppose card $B_{(2,3)} = k^{m-2}$, and consider B' as defined above. Then card $B'_{(m-1,m)} = k^{m-1}$ by Proposition 1.4. Hence, B' satisfies the assumptions of Proposition 1.1 and Proposition 1.3, thus, by Proposition 1.3, card $B'_{(1,2)} = k^{m-1}$. But, by Proposition 1.4, card $B'_{(1,2)} = k^{m-2}$. Hence, it cannot hold $B_{(2,3)} = k^{m-2}$, thus the proof is complete. ▲

As a consequence we obtain:

<u>Proposition 1.6:</u> Let $m \geq 4$; then $\underline{B}$ is totally reflexive.

<u>Proposition 1.7:</u> Let $m \geq 4$; then $\underline{B}$ is totally reflexive and totally symmetric, and $k^{m-1} <$ card $B < k^m$.

<u>Proof:</u> Let $\underline{B}^* = \langle B^*,F\rangle$ be the subalgebra of $\underline{A}^m$ generated by ι_m(see 1.1). Since ι_m generates $\underline{B}^*$and ι_m is totally symmetric, $\underline{B}^*$ is totally reflexive and totally symmetric. By proposition 1.6 we have $B^* \subseteq B$ since $\iota_m \subseteq B$, therefore card $B^* \leq$ card $B < k^m$. Since $m \geq 4$, k^{m-1} = card $B_{(1,2)}$, and $B_{(1,2)}$ is a proper subset of ι_m. Hence, $k^{m-1} <$ card B^*. ▲

<u>Result:</u> Without loss of generality, we may assume that for $m \geq 4$, $\underline{B}$ is totally reflexive and totally symmetric.

II. The cases m = 2 and m = 3

Let $m = 2$. For $h \geq 3$, we assume that ι_h generates $\underline{A}^h$. The case if ι_h does not generate $\underline{A}^h$ will be dealt in the last section. For $C \subseteq A^2$, set $C^{-1} = \{(a,b) \mid (b,a) \in C\}$. For $C_1,C_2 \subseteq A^2$, let $C_1 \circ C_2 = \{(a,c) \mid (a,b) \in C_1$ and $(b,c) \in C_2$ for some $b \in A\}$ be the relation product of C_1 and C_2. Let $\Delta(A)$ be the diagonal of A^2. Since $\underline{A}$ has no proper subalgebra, either $\Delta(A) \subseteq B$ or $B \cap \Delta(A) = \emptyset$. In the first case B is reflexive, and in the second case areflexive. We say that $z \in A$ is a zero of a relation $\varrho \subseteq A^2$ if $(z,a) \in \varrho$ for all $a \in A$ and that $u \in A$ is a unit of $\varrho \subseteq A^2$ if $(a,u) \in \varrho$ for all $a \in A$. If B is areflexive, then B has neither a zero nor a unit.

<u>Proposition 2.1:</u> Let $B \circ B^{-1} = A^2$, then B has at least one zero and at least one unit.

<u>Proof:</u> It suffices to prove that B has a unit. In fact, B has then a zero since $B \circ B^{-1} = A^2$ implies that $B^{-1} \circ B = A^2$(i.e. B^{-1} has a unit). For $2 \leq i \leq k$, let $C_i = \{(a_1,\ldots,a_i) \mid \text{for some } b \in A,\ (a_j,b) \in B \text{ for } 1 \leq j \leq i\}$. Notice that $C_2 = B \circ B^{-1} = A^2$ and that $\langle C_i,F\rangle$ is a totally symmetric subalgebra of $\underline{A}^i$. But $C_i = A^i$ implies that C_{i+1} is totally reflexive and hence equals A^{i+1}. Thus, by induction, it holds $C_k = A^k$. But then $(0,\ldots,k-1) \in C_k$ and so, for some $u \in A$ we have $(a,u) \in B$ for all $a \in A$, i.e. u is a unit of B. ▲

If B is areflexive, then B can be replaced by the reflexive relation $B \circ B^{-1}$. In fact, in this case B has no unit. Hence, by Proposition 2.1 $\langle B \circ B^{-1},F\rangle$ is a proper subalgebra of $\underline{A}^2$ and card $(B \circ B^{-1}) > k$. Thus, we can assume that B is reflexive. If $B \wedge B^{-1} = \Delta(A)$, then B is antisymmetric. Otherwise $B \wedge B^{-1}$ properly contains $\Delta(A)$. In this latter case, we can also assume that B is symmetric (we can replace B by $B \wedge B^{-1}$ which is symmetric). This gives a totally reflexive and totally symmetric relation B which is considered in the last section. Hence, we can assume that B is antisymmetric and that no proper subalgebra of $\underline{A}^2$ properly containing $\Delta(A)$ is symmetric. Since B is antisymmetric, it has at most one zero and at most one unit. We have $B \circ B^{-1} = A^2$ since $B \circ B^{-1}$ is symmetric and properly contains $\Delta(A)$. Thus, by Proposition 2.1, B has exactly one zero and exactly one unit.

<u>Proposition 2.2:</u> If B has a unit u, and is antisymmetric, then $B \circ B \neq A^2$.

<u>Proof:</u> Let $a \in A$; if $(u,a) \in B \circ B$, then $(u,b) \in B$ and $(b,a) \in B$ for some $b \in A$. But $(u,b) \in B$ gives $b = u$ since $(b,u) \in B$ and since B is antisymmetric. Hence, $(u,a) \in B$ and so $u = a$. Thus $(u,a) \notin B \circ B$ for $a \neq u$. ▲

Now let us assume that $\underline{B}$ is maximal among the antisymmetric subalgebras of $\underline{A}^2$. Consider B o B; $\langle B \circ B, F\rangle$ is a subalgebra of $\underline{A}^2$. By Proposition 2.2 we obtain $B \circ B \neq A^2$. If B o B is not antisymmetric, then $(B \circ B) \wedge (B^{-1} \circ B^{-1})$ is symmetric and lies properly between $\Delta(A)$ and A^2, contrary to our assumption on $\underline{A}^2$. Hence, B o B is antisymmetric, therefore B o B = B by the maximality of B. Thus B is transitive, consequently, it is a partial order with a zero and a unit, i.e. it falls into the class O of Rosenberg's classification. Except for B being totally reflexive and totally symmetric, we have eliminated the case m=2.

Let m = 3. We assume that $\underline{A}^h$ has no proper totally reflexive and totally symmetric subalgebra for $h \geq 3$. Let $\Delta_3(A) = \{(a,a,a) | a \in A\}$. Then it holds

<u>Proposition 2.3:</u> $B_{(1,2,3)} = \Delta_3(A)$

<u>Proof:</u> By proposition 1.1 it holds card $p_3(B) = k^2$. Hence card $B_{(1,2)}$=k or k^2. If card $B_{(1,2)} = k^2$, then it follows card $B_{(1,2,3)}$=k, i.e. card $B_{(1,2,3)} = \Delta_3(A)$. If card $B_{(1,2)} = k^2$, then it follows card $B_{(1,2,3)}$ = k, then $\langle p_1(B_{(1,2)}), F\rangle$ is a subalgebra of $\underline{A}^2$ and therefore it is an automorphism, say s, of $\underline{A}$. Then s is either the identity or all cycles of some power of s have the same prime length since $\underline{A}$ has no proper subalgebras, powers of automorphisms are automorphisms and since the fixed points of an automorphism form a subalgebra. But then s must be the identity since otherwise B falls into the class P of Rosenberg's classification. Hence, $B_{(1,2)} = B_{(1,2,3)} = \Delta_3(A)$. ▲

We consider the subalgebra $\langle B', F\rangle$ of $\underline{A}^3$ defined in the first part by $B' = \{(a_2, a_3, a_3') \,|\, (a_1, a_2, a_3), (a_1, a_2, a_3') \in B$ for some $a_1 \in A\}$. $\langle B', F\rangle$ contains $\Delta_3(A)$. Applying Proposition 1.1 to B, we see that card $B'_{(2,3)} = k^2$. Hence, $B' \wedge l_3 \neq \Delta_3(A)$. If $B \wedge l_3 = \Delta_3(A)$, then card $B'_{(1,2)} = k$, and therefore $B' \neq A^3$ since card $B > k^2$, card $p_3(B) = k$ and card $B' > k^2$. Thus, replacing B by B' if necessary, we can assume that $B \wedge l_3 = \Delta_3(A)$. Up to symmetry, there are exactly two possibilities:

(i) card $B_{(1,2)} = k^2$, card $B_{(1,3)}$ = card $B_{(2,3)}$ = k,

(ii) card $B_{(1,2)}$ = card $B_{(1,3)} = k^2$, card $B_{(2,3)}$ = k.

In the first case we define $B_i \subseteq A^i$ for $i \geq 2$ by $B_i = \{(a_1,\ldots,a_i) \mid$ for some $a,b \in A$, $(a_j,a,b) \in B$ for $1 \leq j \leq i\}$. $\langle B_i,F\rangle$ is a totally symmetric subalgebra of $\underline{A}^i$.

<u>Proposition 2.4:</u> Case (i) cannot occur.

<u>Proof:</u> Since card $B_{(1,2)} = k^2$ we have $\Delta(A) \subseteq B_2$. Since card $B > k^2$, there are distinct $a,b,c \in A$ such that $(a,b,c) \in B$. Since $(b,b,c) \in B$, we have $(a,b) \in B_2$. Thus, $B_2 = A^2$. Now, $B_n = A^n$ obviously implies that B_{n+1} is totally reflexive, and hence $B_{n+1} = A^{n+1}$. Thus, $B_k = A^k$, and therefore $(0,\ldots,k-1) \in B_k$. Therefore, there are $a,b \in A$ such that $(j,a,b) \in B$ for $0 \leq j \leq k-1$. For $j = b$, this states that $(b,a,b) \in B$, i.e. $b = a$ since card $B_{(1,3)} =$ card A. But then $(j,a,a) \in B$ for $j \neq a$, which contradicts card $B_{(2,3)} = k$. Thus, case (i) cannot occur. ▲

Next, we show that case (ii) cannot occur. Let $B^{(1)} = \{(b,a,c) \mid (a,b,c) \in B\}$ and consider $B \cap B^{(1)}$. Now, $(a,b,b) \in B^{(1)}$ implies $a = b$, therefore $B \cap B^{(1)} \cap l_3 = \{(a,a,b) \mid a,b \in A\}$. If $B \cap B^{(1)} \neq B \cap B^{(1)} \cap l_3$, then, by replacing B by $B \cap B^{(1)}$, we have case (i) which we have just shown to be impossible. Thus, $B \cap B^{(1)} = \{(a,a,b) \mid a,b \in A\}$ i.e. if (a,b,c), $(b,a,c) \in B$, then $a = b$.

For a partition P on $\{1,\ldots,n\}$ let $\Delta P \subseteq A^n$ be defined by $\Delta P = \{(a_1,\ldots,a_n) \mid a_i = a_j$ if i and j are in the same P-class$\}$. Let $\langle C,F\rangle$ be the subalgebra of $\underline{A}^4$ generated by $\Delta_{(1,2)(3,4)} \cup \Delta_{(1,3)(2,4)} \cup \Delta_{(1,4)(2,3)}$: C is totally symmetric. Obviously, $p_i(C) \supseteq l_3$, and therefore $p_i(C) = A^3$ for $1 \leq i \leq 4$.

<u>Proposition 2.5:</u> For $\{i,j,r,s\} = \{1,2,3,4\}$, we have $C_{(i,j)} = C_{(i,j)(r,s)}$.

<u>Proof:</u> Consider $B^{(2)} = \{(a_1,a_2,a_3,a_4) \mid (a_1,a_2,a_5),(a_5,a_3,a_4),(a_2,a_1,a_6),(a_6,a_3,a_4) \in B$ for some $a_5,a_6 \in A\}$. $\langle B^{(2)},F\rangle$ is a subalgebra of $\underline{A}^4$. One easily checks that $C \subseteq B^{(2)}$. Note next that if $(a_1,a_2,a_3,a_3) \in B^{(2)}$, then $a_6 = a_5 = a_3$ and therefore $(a_1,a_2,a_3),(a_2,a_1,a_3) \in B$; hence $a_1 = a_2$. Since $C \subseteq B^{(2)}$, this property also holds for C. Since C is totally symmetric, for $\{i,j,r,s\} = \{1,2,3,4\}$ we have $C_{(i,j)} = C_{(i,j)(r,s)}$. ▲

<u>Proposition 2.6:</u> For all $a_2,a_3,a_4 \in A$ there is a unique $a_1 \in A$ such that

$(a_1,a_2,a_3,a_4) \in C$.

Proof: Since $p_1(C) = A^3$, for all $a_2,a_3,a_4 \in A$ there is an $a_1 \in A$ such that $(a_1,a_2,a_3,a_4) \in C$. We next show that a_1 is unique. Consider $B^{(3)} = \{(a_1,a_2,a_3) \mid (a_1,a_3,a_4,a_5),(a_2,a_3,a_4,a_5) \in C$ for some $a_4,a_5 \in A\}$. $\langle B^{(3)},F\rangle$ is a subalgebra of $\underline{A}^3$. Obviously, card $B^{(3)}_{(1,2)} = k^2$. If $(a_1,a_2,a_2) \in B^{(3)}$, then $(a_2,a_2,a_4,a_5) \in C$, therefore we have $a_4 = a_5$; hence, $(a_1,a_2,a_4,a_4) \in C$ and $a_1 = a_2$ by Proposition 2.5. Thus card $B^{(3)}_{(2,3)} = k$; similarly, card $B^{(3)}_{(1,3)} = k$. Since case (i) is impossible, we must have $B^{(3)} = B^{(3)}_{(1,2)}$. But $B^{(3)} = B^{(3)}_{(1,2)}$ means that the following implication is fulfilled: If $(a_1,a_2,a_3,a_4),(a_1',a_2,a_3,a_4) \in C$, then $a_1=a_1'$. ▲

Thus $a_1 = l(a_2,a_3,a_4)$ defines a function $l:A^3 \longrightarrow A$. Obviously, $l(a,b,c) = l(a,c,b) = l(b,c,a)$; $l(a,a,b) = b$; and $l(a,b,l(a,b,c)) = c$.

Proposition 2.7: $l(l(a,b,c),b,d) = l(a,b,l(c,b,d))$ for all $a,b,c,d \in A$.

Proof: Consider $C' \subseteq A^4$ defined by

$C' = \{(a_1,a_2,a_3,a_4) \mid l(l(a_1,a_2,a_3),a_2,a_4) = l(a_1,a_2,l(a_3,a_2,a_4))\}$;

$\langle C',F\rangle$ is a subalgebra of $\underline{A}^4$. Using the identities mentioned above, it is easy to show that for $1 \le i < j \le 4$, card $C'_{(i,j)} = k^3$. Hence, $C' = A^4$, i.e. $l(l(a,b,c),b,d) = l(a,b,l(c,b,d))$ for all $a,b,c,d \in A$. ▲

Proposition 2.8: $l(a,b,c) = l(l(a,d,c),d,b)$ for all $a,b,c,d \in A$.

Proof: Let $C'' \subseteq A^4$ be defined by

$C'' = \{(a_1,a_2,a_3,a_4) \mid l(a_1,a_2,a_3) = l(l(a_1,a_4,a_3),a_4,a_2)\}$

$\langle C'',F\rangle$ is a subalgebra of $\underline{A}^4$. Again, the above equations imply that C'' is totally reflexive and hence $C'' = A^4$. Thus, $l(a,b,c) = l(l(a,d,c),d,b)$ for all $a,b,c,d \in A$. ▲

From these equations it follows that $\langle A,l\rangle$ is isomorphic to the n-th power of the two-element algebra $\underline{L_4}$ for some n (Denecke [1]). Therefore, all term functions of A satisfy $f(\underline{x} + \underline{y} + \underline{z}) = f(\underline{x}) + f(\underline{y}) + f(\underline{z})$, where $\underline{x}, \underline{y}, \underline{z} \in A^j$ if f is j-ary and f is necessarily of the form $x_1 + \dots + x_j + a$ where $a \in A$.

Let us consider $\tilde{B}$ from card $B_{(1,2)} =$ card $B_{(1,3)} = k^2$ and card $B_{(2,3)}$

$= k$ it follows $B \cap \Delta_3 = \Delta_{(1,2)} \cup \Delta_{(1,3)}$. Given the above characterization of the term functions of $\underline{A}$, this can only happen if all term functions of $\underline{A}$ are unary. If card $A \geqslant 3$, then $\langle \iota_3, F\rangle$ is a proper subalgebra of $\underline{A}^3$, contrary to our assumption that ι_3 generates $\underline{A}^3$. Finally, if $k = 2$, then the unary term functions of $\underline{A}$ are all four unary functions on A. But then $C = \{(a_1,a_2,a_3,a_4) | a_1 + a_2 = a_3 + a_4 \pmod 2\}$, i.e. it falls into the class L of Rosenberg's classification. Hence, the case (ii) cannot occur.

Result: Except for B being totally reflexive and totally symmetric, we have eliminated the cases $m = 2$ and $m = 3$, i.e. now we have to consider the case of totally reflexive and totally symmetric relations.

III. The totally reflexive and totally symmetric case

Let $B \subseteq A^m$ be totally reflexive and totally symmetric. Note that this implies that $m \leqslant k$ since otherwise $B = A^m$. We choose $h \leqslant k$ maximal with respect to the property that $\underline{A}^h$ contains a proper totally symmetric and totally reflexive subalgebra, say $\underline{C} = \langle C, F\rangle$. We also assume without loss of generality that there is no totally reflexive and totally symmetric algebra strictly between $\underline{C}$ and $\underline{A}^h$.

For $h \leqslant n \leqslant k$ let C_n be the subset of all $(a_1,\dots,a_n) \in A^n$ for which there is an $a \in A$ such that $(a_{i_1},\dots,a_{i_{h-1}},a) \in C$ for each $(h-1)$-element subset $\{i_1,\dots,i_{h-1}\}$ of $\{1,\dots,n\}$. $\langle C_n, F\rangle$ is a subalgebra of $\underline{A}^n$ and is totally symmetric since C is.

Proposition 3.1: Either $C = C_h$ or $C_h = A^h$.

Proof: If $(a_1,\dots,a_h) \in C$, then choose $a = a_1$. By total reflexivity and total symmetry of C $(a_1,\dots,a_{i-1},a_{i+1},\dots,a_h,a_1) \in C$ for $1 \leqslant i \leqslant h$; hence $(a_1,\dots,a_h) \in C_h$. Thus, $C \subseteq C_h$. By maximality of C, we must either have $C = C_h$ or $C_h = A^h$.

Proposition 3.2: If $C_h = A^h$, then $C_k = A^k$.

Proof: We know that C_n is totally symmetric for $h \leqslant n \leqslant k$. By maximality of h, $C_n = A^n$ if C_n is totally reflexive. But it follows immediately from the definition of C_n that C_{n+1} is totally reflexive if $C_n = A^n$. Thus, by induction, Proposition 3.1 implies $C_k = A^k$. ▲

Suppose that $C_k = A^k$; thus $(0,\dots,k-1) \in C_k$. Therefore, there is an $a \in A$ such that $(i_1,\dots,i_{h-1},a) \in C$ for any (h-1)-element subset $\{i_1,\dots,i_{h-1}\}$ of $A = \{0,\dots,k-1\}$. Together with the total symmetry and total reflexivity of C, this means that a is in the center of C. Hence, C is central and falls into the class C of Rosenberg's classification. Thus, we can assume $C_k = A^k$ by the Propositions 3.1 and 3.2 we must have $C_h = C$, such a C is called homogeneous. For h = 2, if $(a_1,a_3),(a_2,a_3) \in C$, then $(a_1,a_2) \in C$, i.e. C is transitive (using symmetry). Hence, C is a non-trivial equivalence relation and falls into the class E of Rosenberg's classification. Thus, we can assume that $h \geq 3$.

C is called strongly homogeneous if any $(a_1,\dots,a_h) \in A^h$ belongs to C whenever there is $(v_1,\dots,v_h) \in C$ such that $(a_1,\dots,a_{i-1},a_{i+1},\dots,a_h,v_j) \in C$ for all $1 \leq i, j \leq h$ with $i \neq j$. We shall use the maximality of C in order to prove that C is strongly homogeneous.

For $h \leq r \leq k$, let $C_r \subseteq A^r$ be defined by

$C_r = \{(a_1,\dots,a_r) | \pi_E(a_1,\dots,a_r) \in C$ for every $E \subseteq \{1,\dots,r\}$ with card $E = h\}$.

$\langle C_r,F\rangle$ is a subalgebra of $\underline{A}^r$. Notice that $C_h = C$ for $r > h$, C_r is totally symmetric but not totally reflexive. For $h \leq r \leq k$ define $D_r \subseteq A^r$ to be the set of all $(a_1,\dots,a_r)$ for which there is $(b_1,\dots,b_r) \in C_r$ such that $(a_{i_1},\dots,a_{i_{h-2}},a_j,b_j) \in C$ for every $1 \leq j \leq r$ and every $\{i_1,\dots,i_{h-2}\} \subseteq \{1,\dots,j-1,j+1,\dots,r\}$. $\langle D_r,F\rangle$ is a subalgebra of $\underline{A}^r$. The total symmetry of C and C_r imply that D_r is totally symmetric. Notice that C is strongly homogeneous if and only if $C = D_h$.

<u>Proposition 3.3:</u> Either $C = D_h$ or $D_h = A^h$.

<u>Proof:</u> Let $(a_1,\dots,a_h) \in C$. We can choose $b_i = a_i$ for $1 \leq i \leq h$ since $C = C_h$. The total reflexivity of C implies that $(a_1,\dots,a_h) \in D_h$. Thus, $C \subseteq D_h$, and therefore $C = D_h$ or $D_h = A^h$ by the maximality of C. ▲

Proposition 3.4: If $D_h = A^h$, then $D_k = A^k$.

Proof: Suppose $D_h = A^h$. If we can show that $D_n = A^n$ implies the total reflexivity of D_{n+1}, then the maximality of h will imply $D_{n+1} = A^{n+1}$. Hence, by induction, we shall get $D_k = A^k$. If $D_n = A^n$, then for any $a_1,\ldots,a_n \in A$ there are $(b_1,\ldots,b_n) \in C_n$ such that $(a_{i_1},\ldots,a_{i_{h-1}},a_j,b_j) \in C$ for every $\{i_1,\ldots,i_{h-2}\} \subseteq \{2,\ldots,j-1,j+1,\ldots,n\}$. But $(b_1,\ldots,b_n) \in C_n$ implies $(b_1,b_1,\ldots,b_n) \in C_{n+1}$, hence $(a_1,a_1,\ldots,a_n) \in D_{n+1}$. This proves that D_{n+1} is totally reflexive.

If $D_k = A^k$, then $(a_1,\ldots,a_k) \in D_k$ where $\{a_1,\ldots,a_k\} = A$. Thus, there are $(b_1,\ldots,b_k) \in C_k$ such that $(a_{i_1},\ldots,a_{i_{h-1}},a_j,b_j) \in C$ for every $1 \leq j \leq k$ and every $\{i_1,\ldots,i_{h-2}\} \subseteq \{1,\ldots,j-1,j+1,\ldots,k\}$. Since C is totally reflexive, this means that $(c_1,\ldots,c_{h-2},a_j,b_j) \in C$ for all $c_1,\ldots,c_{h-1} \in A$ and all $1 \leq j \leq k$.

Proposition 3.5: C is strongly homogeneous.

Proof: We must show $C = D_h$. By propositions 3.3 and 3.4 it suffices to show $D_k \neq A^k$. Let $\{a_1,\ldots,a_k\} = A$; if $D_k = A^k$, then $(a_1,\ldots,a_k) \in D_k$, via, say, $(b_1,\ldots,b_k) \in C_k$ (see definition of D_k). Thus, $(c_1,\ldots,c_{h-2}, a_j,b_j) \in C$ for all $c_1,\ldots,c_{h-1} \in A$ and all $1 \leq j \leq k$. We prove by induction on n that $(a_1,\ldots,a_n,b_{n+1},\ldots,b_n) \in C$ for $0 \leq n \leq h$. Since $(b_1,\ldots,b_k) \in C_k$, $(b_1,\ldots,b_h) \in C_h = C$. Thus the case $n = 0$ is proved. Suppose that for $n < h$, $(a_1,\ldots,a_n,b_{n+1},\ldots,b_h) \in C$. In order to show $(a_1,\ldots,a_{n+1},b_{n+2}, \ldots,b_h) \in C$, we shall use the homogeneity of C. In $(a_1,\ldots,a_{n+1},b_{n+2}, \ldots,b_h)$, if we replace every element distinct from a_{n+1} by b_{n+1}, then this resulting h-tuple contains both a_{n+1} and b_{n+1}, and so, as noted above, it is contained in C. On the other hand, replacing a_{n+1} by b_{n+1} gives us an element of C by our induction hypothesis. Hence, by the homogeneity of C, $(a_1,\ldots,a_{n+1},b_{n+2},\ldots b_h) \in C$. Thus, by induction, $(a_1,\ldots,a_h) \in C$. But $\{a_1,\ldots,a_k\} = A$; therefore $\{a_1,\ldots,a_h\}$ is an arbitrary h-element subset of A. Since C is totally reflexive, this implies $C = A^h$, this contradicts our assumption. Hence, $C_k = A^k$, and thus, $C_h = C$.

To construct a contradiction, now we shall show that C is h-universal (see 4.1), i.e. we must find a surjection $\Phi: A \longrightarrow \{0,\dots,h-1\}^m$ such that $C = \Phi^{-1}(\eta)$ where η is the m-fold power of the primitive h-elementary relation

Our first step is to find ker(Φ), the equivalence relation on A induced by Φ. Let $E = \{(a,b) \mid (a_1,\dots,a_{h-2},a,b) \in C$ for all $a_1,\dots,a_{h-2} \in A\}$. E is reflexive and symmetric. Let $(a,b), (b,c) \in E$. Then, using b as a replacement element, the homogeneity of C implies that for all $a_1,\dots,a_{h-2} \in A$, $(a_1,\dots,a_{h-2},a,c) \in C$. Thus, E is an equivalence relation; let it have q equivalence classes. We may assume that $A' = \{0,\dots,q-1\}$ contains one element from each E-class. Let $F: A \longrightarrow A'$ be the function defined by $(a,F(a)) \in E$ for all $a \in A$. Let $\tilde{C} \subseteq A^h$ be defined by $\tilde{C} = \{(a_1,\dots,a_h) \mid (F(a_1),\dots,F(a_h)) \in C\}$.

<u>Proposition 3.6:</u> Let $(a,b) \in E$ and $a_1,\dots,a_{h-1} \in A$, then $(a_1,\dots,a_{h-1},a) \in C$ if and only if $(a_1,\dots,a_{h-1},b) \in C$.

<u>Proof:</u> Let $(a_1,\dots,a_{h-1},b) \in C$. Then $(a_1,\dots,a_{h-1},a) \in C$ follows from $(a,b) \in E$ and from homogeneity of C using b as a replacement element.

<u>Proposition 3.7:</u> $\tilde{C} = C$.

<u>Proof:</u> Let $(a_1,\dots,a_h) \in C$. By Proposition 3.6 $(F(a_1),a_2,\dots,a_h)$, $(F(a_1),F(a_2),\dots,a_h),\dots,(F(a_1),\dots,F(a_h)) \in C$. Hence, $(a_1,\dots,a_h) \in \tilde{C}$. Conversely, let $(a_1,\dots,a_h) \in \tilde{C}$, thus $(F(a_1),\dots,F(a_h)) \in C$. By Proposition 3.6 we have $(a_1,F(a_2),\dots,F(a_h)),\dots,(a_1,\dots,a_h) \in C$.

Let $D = C \cap (A')^h$; D is totally symmetric and totally reflexive. Proposition 3.6 implies that, in the definition of the strong homogeneity, we can replace $(v_1,\dots,v_h)$ by $(F(v_1),\dots,F(v_h))$; hence, D is strongly homogeneous.

Now, for $h \leq i \leq k$, let $C_i \subseteq A^i$ be defined by $(a_1,\dots,a_i) \in C_i$ if and only if there is an $f: \{0,\dots,h-1\}^{h^i} \longrightarrow A$ such that $f(e_j^{\,i}) = a_j$ for $1 \leq j \leq i$ and $(f(b_1),\dots,f(b_h)) \in C$ for all $(b_1,\dots,b_h) \in \eta_i$. If $m = h^i$, then we shall denote η by η_i.

Proposition 3.8: $C_h = A^h$.

Proof: Let $(a_1,\dots,a_h)$ be an arbitrary element of A^h. Let $\tilde{z} = (0,\dots,h-1)$ and let $w\colon \{0,\dots,h-1\} \to A$ be defined by $w(j) = a_{j+1}$ for $0 \leq j \leq h-1$. Define $f_w\colon \{0,\dots,h-1\}^{h^h} \to A$ by $f_w(b) = w(b(\tilde{z}))$. Then $f_w(e_j{}^h) = w(e_j{}^i(\tilde{z})) = w(j-1) = a_j$ for $1 \leq j \leq h$. Moreover, if $(b_1,\dots,b_h) \in \eta_h$, then $(f_w(b_1),\dots,f_w(b_h)) = w(b_1(\tilde{z})),\dots,w(b_h(\tilde{z}))) \in \iota_h$ since $(b_1(\tilde{z}),\dots,b_h(\tilde{z})) \in \{0,\dots,h-1\}_h$, where $\{0,\dots,h-1\}_h \subset \{0,\dots,h-1\}^h$ is defined by $\{0,\dots,h-1\}_h = \{(a_1,\dots,a_h) \mid a_i = a_j$ for some $i \neq j$ and $a_1,\dots,a_h \in \{0,\dots,h-1\}\}$. Thus $(a_1,\dots,a_h) \in C_h$. ▲

Proposition 3.9: For $h < i < k$, $\langle C_i,F\rangle$ is a totally symmetric subalgebra of $\underline{A}^i$.

Proof: Let g be an n-ary term function of the algebra $\underline{A}$ and for $1 \leq j \leq n$, let $\underline{a}_j \in C_i$ via the function $f_j\colon \{0,\dots,h-1\}^{h^i} \to A$ (see definition of C_i). Let $f = g(f_1,\dots,f_h)$ and let $g(\underline{a}_1,\dots,\underline{a}_n) = (g_1,\dots,g_i)$, i.e. $g_j = g(a_{1j},\dots,a_{nj})$ where $\underline{a}_p = (a_{p1},\dots,a_{pi})$. Then $f(e_j{}^i) = g(f_1(e_j{}^i,\dots, f_n(e_j{}^i)) = g(a_{1j},\dots,a_{nj}) = g_j$. If $(b_1,\dots,b_h) \in \eta_i$, then $(f(b_1),\dots,f(b_h)) = (g(f_1(b_1),\dots,f_n(b_1)),\dots,(g(f_1(b_h),\dots,f_h(b_h))) \in C$ since $(f_j(b_1),\dots,f_j(b_h)) \in C$ for $1 \leq j \leq n$ and since $\underline{C}$ is a subalgebra of $\underline{A}^h$. Thus, $(g_1,\dots,g_i) \in C_i$, and so $\langle C_i,F\rangle$ is a subalgebra of $\underline{A}^i$. To show total symmetry let s be a permutation of $\{1,\dots,i\}$ and let $(a_1,\dots,a_i) \in C_i$ via f. Then $(a_{s(1)},\dots,a_{s(i)}) \in C_i$ via f_s, where $f_s(b(x_1,\dots,x_i)) = f(b(x_{s(1)},\dots,x_{s(i)}))$. ▲

Proposition 3.10: There is a function $f\colon \{0,\dots,h-1\}^h \to A$ such that $f(\eta_j{}^k) = j-1$ for $1 \leq j \leq k$ and such that $(f(b_1),\dots,f(b_h)) \in C$ for all $(b_1,\dots,b_h) \in \eta_k$.

Proof: We first prove $C_k = A^k$ by induction. By Proposition 3.8 we have $C_h = A^h$. Assume $C_n = A^n$, to show that $C_{n+1} = A^{n+1}$, due to Proposition 3.9 it suffices to show that C_{n+1} is totally reflexive. For this we need only show that $(a_1,\dots,a_n) \in A^n$ implies $(a_1,a_1,a_2,\dots,a_n) \in C_{n+1}$. Thus, let $(a_1,\dots,a_n) \in A^n = C_n$ via f. Define $\tilde{f}\colon \{0,\dots,h-1\}^{h^{n+1}} \to A$ so that for $b\colon \{0,\dots,h-1\}^{n+1} \to \{0,\dots,h-1\}$, $f(b(x_1,\dots,x_{n-1})) =$

$f(b(x_2,x_2,\dots,x_{n+1}))$. Obviously, we have $(a_1,a_1,a_2,\dots,a_n)\in C_{n+1}$ via $\tilde{f}$. Thus, $C_k = A^k$, and therefore $(0,\dots,k-1)\in C_k = A^k$, say via f'. ▲

We now fix $f'\colon \{0,\dots,h-1\}^{h^k} \longrightarrow A$ such that C has the property described in Proposition 3.10. Note that Proposition 3.10 is satisfied if C is substituted by D as shown by the function f = F(f'). D is not an algebra under the term functions of $\underline{A}$ since $\underline{A}'$ is not closed under the term functions of $\underline{A}$. We shall prove that q = card A' = h^m for some m. Notice that D has the following property:

(+) a = b for $a,b\in A'$ if $(a_1,\dots,a_{h-2},a,b)\in D$ for all $a_1,\dots,a_{h-2}\in A'$.

Let $s,t\in\{0,\dots,h-1\}^k$, $j\in\{0,\dots,h-1\}$, $g\colon\{0,\dots,h-1\}^k\to\{0,\dots,h-1\}$. We define $g_j^{\,t}(s)$ to be g(s) if s ≠ t and to be j, otherwise. Define $B_j^{\,t}$ to be $\{b\colon\{0,\dots,h-1\}^k\longrightarrow\{0,\dots,h-1\} \mid b(t) = j\}$.

<u>Proposition 3.11:</u> Let $(a_1,\dots,a_h)\in\iota_h$, $t\in\{0,\dots,h-1\}^k$, $g\colon\{0,\dots,h-1\}^k\longrightarrow\{0,\dots,h-1\}$, $b_i\in B_{a_i}^{\,t}$ for $1\le i\le h-2$. Then $(f(b_1),\dots,f(b_{h-2}),f(g_{a_{h-1}}^{\,t}),f(g_{a_h}^{\,t}))\in D$.

<u>Proof:</u> Since D is universal via f, it suffices to show that $(b_1,\dots,b_{h-2},g_{a_{h-1}}^{\,t},g_{a_h}^{\,t})\in\iota_k$. Consider the t'th component of this h-tuple for $t'\in\{0,\dots,h-1\}^k$. If t' = t, then we obtain $(b_1(t'),\dots,b_{h-2}(t'),g(t'),g(t'))\in\iota_h$. Thus, $(b_1,\dots,b_{h-2},g_{a_{h-1}}^{\,t},g_{a_h}^{\,t})\in\iota_k$ ▲

<u>Proposition 3.12:</u> Let $g\colon\{0,\dots,h-1\}^k\longrightarrow\{0,\dots,h-1\}$, $t\in\{0,\dots,h-1\}^k$, $b_p\colon\{0,\dots,h-1\}^k\longrightarrow\{0,\dots,h-1\}$ for $1\le p\le h$, and let $(f(g_0^t),\dots,f(g_{h-1}^t))\in D$. Consider $(f(b_1),\dots,f(b_h))$: if, for $1\le r<s\le h$, we replace the rth component, $f(b_r)$, by $f(g_{r-1}^t)$ and replace the sth component, $f(b_s)$, by $f(g_{s-1}^t)$, then we get a member of D.

<u>Proof:</u> It suffices to take r = h-1 and s = h. If $(b_1(t),\dots,b_{h-2}(t),h-2,h-1)\in\iota_h$, then the result follows from Proposition 3.11. Otherwise, we can assume that $b_i(t) = i-1$ for $1\le i\le h-2$. We shall prove by induction that for $0\le n\le h-2$, $(f(b_1),\dots,f(b_n),f(g_n^{\,t}),\dots,f(g_{h-1}^t))\in D$. For n = 0, this is an assumption of the Proposition 3.12. Suppose that, for $n<h-2$, we have $(f(b_1),\dots,f(b_n),f(g_n^{\,t}),\dots,f(g_{h-1}^t))\in D$. Then, using

$f(g_n^t)$ as the replacement element, that $(f(b_1),\dots,f(b_n),f(b_{n+1}),f(g_{n+1}^t)$ $\dots,f(g_{h-1}^t))\in D$. To see this, consider $(b_1,\dots,b_n,b_{n+1},g_{n+1}^t,\dots,g_{h-1}^t)$: replace any component distinct from b_{n+1} by g_n^t. Then $g_n^t(t) = n = b_{n+1}(t)$, while $g_n^t(t') = g_{h-2}^t(t') = g_{h-1}^t(t') = g(t')$ for $t' \neq t$. Thus, each replacement h-tuple belongs to η_k; applying f to each replacement h-tuple, we get a member of D. Now, by induction, $(f(b_1),\dots,f(b_{h-2}),$ $f(g_{h-2}^t),\ f(g_{h-1}^t))\in D$; this proves Proposition 3.12. ▲

<u>Proposition 3.13:</u> With the same assumptions as in Proposition 3.12 we have $f(g_o^t) = \dots = f(g_{h-1}^t)$.

<u>Proof:</u> It suffices to prove that $f(g_{h-2}^t) = f(g_{h-1}^t)$. Proposition 3.12 states that for all $b_i: \{0,\dots,h-1\}^k \longrightarrow \{0,\dots,h-1\}$ where $1 \leq i \leq h-2$, $(f(b_1),\dots,f(b_{h-2}),\ f(g_{h-2}^t),f(g_{h-1}^t))\in D$. Choose $b_1,\dots,b_{h-2}$ from $\{\{0\}^k, \{0,1\}^k,\dots,\{0,\dots,h-1\}^k\}$ and recall that $f(\eta_j^k) = j-1$ for $1 \leq j \leq h$; then we see that $(a_1,\dots,a_{h-2},f(g_{h-2}^t),f(g_{h-1}^t))\in D$ for all $a_1,\dots,a_{h-2}$ $\in l_h$. Thus, we must have $f(g_{h-2}^t) = f(g_{h-1}^t)$ since D has the property (+). ▲

<u>Proposition 3.14:</u> Let $g:\{0,\dots,h-1\}^k \longrightarrow \{0,\dots,h-1\}$, $b_p: \{0,\dots,h-1\}^k$ $\longrightarrow \{0,\dots,h-1\}$ for $1 \leq p \leq h$, $t \in \{0,\dots,h-1\}^k$ and $(f(g_o^t),\dots,f(g_{h-1}^t))\notin D$. Consider $(f(b_1),\dots,f(b_h))$; if for $1 \leq r < s \leq h$ we replace the rth component, $f(b_r)$, by $f(g_{r-1}^t)$ and the sth component, $f(b_s)$, by $f(g_{s-1}^t)$, then we do not receive an element of D.

<u>Proof:</u> It suffices to take $r = h-1$ and $s = h$. For $n < h-2$, consider $(b_1,\dots,b_n,g_n^t,\dots,g_{h-1}^t)$. Replacing any component distinct from g_n^t by b_{n+1}, we get an element of η_k: $b_{n+1}(t) = g_n^t(t) = n$ while for $t' \neq t$, $g_n^t(t') = g_{h-2}^t(t') = g_{h-1}^t(t') = g(t')$. Applying f to each replacement h-tuple yields an element of D. Thus, if $(f(b_1),\dots,f(b_{n+1}),f(g_{n+1}^t),\dots,$ $f(g_{h-1}^t))\in D$, then the homogeneity of D implies that $(f(b_1),\dots,f(b_n),$ $f(g_n^t),\dots,f(g_{h-1}^t))\in D$. Thus, by induction, our assumption $(f(g_o^t),\dots,$ $f(g_{h-1}^t))\notin D$ implies $(f(b_1),\dots,f(b_{h-2}),f(g_{h-2}^t),f(g_{h-1}^t))\notin D$. ▲

<u>Proposition 3.15:</u> Let $t\in \{0,\dots,h-1\}^k$. Then either $f(g_o^t)=\dots=f(g_{h-1}^t)$ or $(f(g_o^t),\dots,f(g_{h-1}^t))\notin D$ for all $g:\{0,\dots,h-1\}^k \longrightarrow \{0,\dots,h-1\}$

<u>Proof:</u> Suppose for $g: \{0,\dots,h-1\}^k \longrightarrow \{0,\dots,h-1\}$, that

$(f(g_0^t),\dots,f(g_{h-1}^t))\in D$; let $\tilde{g}$ $\{0,\dots,h-1\}^k\to\{0,\dots,h-1\}$. By Proposition 3.13 we have $f(g_0^t)=\dots=f(g_{h-1}^t)$. By Proposition 3.12 it holds $(f(\tilde{g}_0{}^t),\dots,f(\tilde{g}_{h-3}^t),f(g_{h-2}^t),f(g_{h-1}^t))\in D$. If $h\geq 4$, then $(f(\tilde{g}_0{}^t),\dots,f(\tilde{g}_{h-1}^t))\in D$ by Proposition 3.14, and therefore $f(\tilde{g}_0{}^t)=\dots=f(\tilde{g}_{h-1}^t)$ by Proposition 3.13. For $h = 3$, notice that $(f(\tilde{g}_0{}^t),f(\tilde{g}_1{}^t),f(g_2{}^t)) = (f(\tilde{g}_0{}^t),f(\tilde{g}_1{}^t),f(g_1{}^t))\in D$ since $f(g_0^t)=f(g_1^t)$ and $(\tilde{g}_0{}^t,\tilde{g}_1{}^t,g_1{}^t)\in\eta_3$. But then, by Proposition 3.14, $(f(\tilde{g}_0{}^t),f(\tilde{g}_1{}^t),f(\tilde{g}_2{}^t))\in D$ and therefore $f(\tilde{g}_0{}^t)=f(\tilde{g}_1{}^t)=f(\tilde{g}_2{}^t)$ by Proposition 3.13. This proves Proposition 3.15.

▲

Let $T = \{t_1,\dots,t_m\}$ be the set of $t\in\{0,\dots,h-1\}^k$ such that $(f(g_0^t),\dots,f(g_{h-1}^t))\notin D$. Let $S = \{0,\dots,h-1\}^k\setminus T$ for g: $\{0,\dots,h-1\}^k \to\{0,\dots,h-1\}$, and let $\hat{g} = (g(t_1),\dots,g(t_m))\in\{0,\dots,h-1\}^m$.

<u>Proposition 3.16:</u> Let g_1,g_2: $\{0,\dots,h-1\}^k\to\{0,\dots,h-1\}$ and $\hat{g}_1 = \hat{g}_2$. Then $f(g_1) = f(g_2)$.

<u>Proof:</u> For all g: $\{0,\dots,h-1\}^k\to\{0,\dots,h-1\}$, $s\in S$ and $0\leq i\leq h-1$, Proposition 3.15 implies that $f(g_i^s)=f(g)$ (if $g(s)=j$, then $g_j^s = g$). Thus, starting from g_1 we may arbitrarily change its value on every member of S without changing its image under f. Since $\hat{g}_1 = \hat{g}_2$, g_1 and g_2 differ only on S. Hence, $f(g_1) = f(g_2)$. ▲

<u>Proposition 3.17:</u> Let g_1,g_2: $\{0,\dots,h-1\}^k\to\{0,\dots,h-1\}$; if $\hat{g}_1 \neq \hat{g}_2$, then $f(g_1) \neq f(g_2)$.

<u>Proof:</u> Let $t\in T$ with $g_1(t)\neq g_2(t)$, we can assume that $g_1(t) = 0$ and $g_2(t) = h-1$. Thus, $(g_1)_0{}^t = g_1$ and $(g_2)_{h-1}^t = g_2$. By Proposition 3.13, $(f((g_1)_0{}^t),\dots,f((g_1)_{h-1}^t))\notin D$; hence Proposition 3.14 implies $(f(g_1),f(g_1)_1{}^t,f(g_2)_2{}^t),\dots,f((g_2)_{h-2}^t),f(g_2))\notin D$. Therefore $f(g_1)\neq f(g_2)$.

▲

<u>Proposition 3.18:</u> card $A' = h^m$.

<u>Proof:</u> We know that $\{f(g)\,|\,g$: $\{0,\dots,h-1\}^k\to\{0,\dots,h-1\}\} = A'$ because $f(e_i^k) = i-1$ for $1\leq i\leq q$. Thus, Propositions 3.16 and 3.17 give card $A' = h^m$.

We define Φ': $A'\to\{0,\dots,h-1\}^m$ as follows: for $a\in A'$, take some g: $\{0,\dots,h-1\}^k\to\{0,\dots,h-1\}$ with $f(g) = a$; then define $\Phi'(a)=\hat{g}$.

By Proposition 3.16 and Proposition 3.17 Φ' is a bijection.

Proposition 3.19: Let $g_1,\dots,g_h: \{0,\dots,h-1\}^{k} \longrightarrow \{0,\dots,h-1\}$, and let $t \in T$ such that $(g_1(t),\dots,g_h(t)) \notin \iota_h$. Then $(f(g_1),\dots,f(g_h)) \notin D$.

Proof: We can assume that $g_i(t) = i-1$ for $1 \le i \le h$. Let $\tilde{o}:\{0,\dots,h-1\}^k \longrightarrow \{0,\dots,h-1\}$ be the zero function: for $s \in \{0,\dots,h-1\}^k$ it holds $\tilde{o}(s) = 0$. Then for $0 \le i,\ j \le h-1$ with $i \neq j$, $(\tilde{o}_0^t,\dots,\tilde{o}_{i-1}^t, g_{j+1}, \tilde{o}_{i+1},\dots, \tilde{o}_{h-1}^t) \in \eta_k$: for $s \neq t$, at s the h-tuple evaluates to $(0,\dots,0,g_{j+1}(s), 0,\dots,0) \in \iota_h$ while at t it evaluates to $(0,1,\dots,i-1,j,i+1,\dots,h-1)$ which is in ι_h since $i \neq j$. Hence, $(f(\tilde{o}_0^t),\dots,f(\tilde{o}_{i-1}^t), f(g_{j+1}), f(\tilde{o}_{i+1}^t), \dots, f(\tilde{o}_{h-1}^t)) \in D$. Thus, if $(f(g_1),\dots,f(g_h)) \in D$, then the strong homogeneity of D implies that $(f(\tilde{o}_0^t),\dots,f(\tilde{o}_{h-1}^t)) \in D$ and therefore $f(\tilde{o}_0^t) = \dots = f(\tilde{o}_{h-1}^t)$ by Proposition 3.13. But $t \in T$ and $\tilde{o}_0^t(t) \neq \tilde{o}_1^t(t)$; hence, by Proposition 3.17 we have $f(\tilde{o}_0^t) \neq f(\tilde{o}_1^t)$. Therefore, we must have $(f(g_1),\dots,f(g_h)) \notin D$.

We complete the proof of 6.4-Theorem 13 by showing that C is h-universal, i.e. $\Phi'(D) = \eta$.

Proposition 3.20: $\Phi'(D) = \eta$.

Proof: Let $(a_1,\dots,a_h) \in (A')^h$, and, for $1 \le i \le h$, choose $g_i: \{0,\dots,h-1\}^k \longrightarrow \{0,\dots,h-1\}$ such that $f(g_i) = a_i$; by Proposition 3.16 we can assume that $g_i(s) = 0$ for all $s \in S$. If $(a_1,\dots,a_h) \in D$, then by Proposition 3.19 we have $(g_1(t),\dots,g_h(t)) \in \iota_h$ for all $t \in T$, i.e. $(\hat{g}_1,\dots,\hat{g}_h) = (\Phi'(a_1),\dots,\Phi'(a_h)) \in \eta$. Conversely, if $(\hat{g}_1,\dots,\hat{g}_h) = (\Phi'(a_1),\dots,\Phi'(a_h)) \in \delta$, then $(g_1(u),\dots,g_h(u)) \in \iota_h$ for all $u \in \{0,\dots,h-1\}^k$, i.e. $(g_1,\dots,g_h) \in \iota_k$.

Hence, by Proposition 3.15 it holds $(f(g_1),\dots,f(g_h)) = (a_1,\dots,a_h) \in D$.

Bibliography

Bacsich, P.D.

[1] Primality and model completions, Algebra Universalis, 3 (1973)

Baker, K.A.

[1] Congruence distributive polynomial reducts of lattices, Algebra Universalis, 9 (1979), 142-145

[2] Finite equational bases for finite algebras in congruence distributive equational classes, Advances in Math., to appear

Baker, K.A. and A.F. Pixley

[1] Polynomial interpolation and the Chinese Remainder Theorem for algebraic systems, Math. Zeitschr. 143 (1975), 165-174

Berman, J.

[1] A new proof of Lyndon's finite basis theorem, Discrete Math. 29 (1980), 229-233

[2] Algebraic properties of k-valued logics, Proceedings of the tenth international symposium on multiple- valued logic, Evanston, Illinois, 1980

Birkhoff, G.

[1] On the structure of abstract algebras, Proc. Cambridge Philosoph. Soc. 31 (1935), 433-454

[2] Lattice theory, Amer. Math. Soc. Colloqu. Publ. vol. XXV rev. ed. Amer. Math. Soc. (1967)

Burle, G.A. (Бурле, Г.А.)

[1] Классы к-значной логики, содержащие все функции одной переменной, дискретный анализ, 10, (1967)

Butler, J.

[1] On complete and independent sets of operations in finite algebras Pac. J. Math. 10 (1960), 1169-1179

Clark, D.M. and P.H. Krauss

[1] Paraprimal algebras, Algebra Universalis, 6 (1976), 165-192

[2] Varieties generated by paraprimal algebras, Algebra Universalis 7 (1977), 93-114

Cohn, P.

[1] Universal Algebra, Harper and Row, Publishers, New York 1965

Csákány, B.

[1] Varieties whose algebras have no idempotent elements, Coll. Math. 2 (1976)

[2] Homogeneous algebras are functionally complete, Algebra Universalis, to appear

Day, A.

[1] Splitting algebras and a weak notion of projectivity, Algebra Universalis, 5 (1975), 153-162

Denecke, K.

[1] Algebraische Fragen nichtklassischer Aussagenkalküle, Doctoral dissertation, Pädagogische Hochschule „Karl Liebknecht", Potsdam, 1978

[2] Schwache Automorphismen präprimaler Algebren, die arithmetische Varietäten erzeugen, Rostock. Math. Kolloqu. 10 (1978), 23-36

[3] Implikative BCK-Algebren, Zeitschr. f. Math. Logik und Grundl. der Math. 6 (1980), 547-554

[4] Präprimale Algebren, die arithmetische Varietäten erzeugen, Algebra and its Applications, Banach Center publications, to appear

[5] Independent sets and free preprimal algebras, Zeitschr. f. Math. Logik und Grundl. der Math., to appear

[6] Die endlichen, subdirekt irreduziblen Elemente der Varietät der beschränkten kommutativen BCK-Algebren, Zeitschr. f. Math. Logik und Grundl. der Math. 27 (1981), 265-272

Dudek, J. and K. Głazek

[1] Some remarks on weak automorphisms, Colloqu. Math. Soc. János Bolyai, 17. Contributions to Universal algebra, Szeged 1975

Eilenberg, S. and M.P. Schützenberger

[1] On pseudovarieties, Advances in Math. 19(1976), 413-418

Foster, A.L.

[1] Generalized „Boolean" theory of universal algebras, Part I, Math. Zeitschr. 58 (1953) , 306-336

[2] Generalized „Boolean" theory of universal algebras, Part II, Math. Zeitschr. 59 (1953), 191-199

[3] The identities of - and unique subdirect factorization within classes of universal algebras, Math. Zeitschr. 62 (1955), 171-188

[4] An existence theorem for functionally complete universal algebras, Math. Zeitschr. 71 (1959), 69-82

[5] On the imbeddability of universal algebras in relation to their identities I, Math. Ann. 138 (1959), 219-238

[6] Functional completeness in the small, algebraic structure theorems and identities, Math. Ann. 143 (1961), 127-146

[7] Generalized equational maximality and primal - in the small-algebras, Math. Zeitschr. 79 (1962), 127-

[8] Congruence relations and functional completeness in universal algebras, structury theorems of hemi-primals I, Math. Zeitschr. 113 (1970), 293-308

Foster, A.L. and A.F. Pixley

[1] Semi-categorical algebras I, semi-primal algebras, Math. Zeitschr. 83 (1964), 147-169

Foster, A.L. and A.F. Pixley

[2] Semi-categorical algebras II, Math. Zeitschr. 85 (1964), 169-184

Fried, E. and E.W.Kiss

[1] Connection between the congruence lattices and polynomial properties, preprint 1979

Fried, E. and A.F.Pixley

[1] The dual discriminator function in Universal algebra, Acta Sci. Math., 41 (1979), 83-100

Frink, O.

[1] The operations of Boolean algebras, Annals of Math., 27 (1925), 477-490

Froemke, J.

[1] A note on functionally complete algebras with no non-trivial subalgebras, Math. Zeitschr. 136 (1974), 353-355

[2] Maximal preprimal clusters, Canad.J. Math. 4 (1975), 746-751

Fuchs, L.

[1] On partially ordered algebras I, Colloqu. Math. 14 (1966), 115-130

Goetz, A.

[1] On weak isomorphisms and weak homomorphisms of abstract algebras, Colloqu. Math. 14 (1966), 163-167

Grätzer, G.

[1] Universal algebra, Springer-Verlag, Berlin-Heidelberg-New York 1979

Grau, A.A.

[1] Ternary Boolean algebra, Bull. Amer. Math. Soc. 53 (1947)

[2] A ternary operation related to the complete disjunction of Boolean algebra, Univ. Nac. Tucuman, Rev.-Ser.A 8 (1951)

Gumm, H.P.

[1] Algebras in permutable varieties; Geometrical properties of affine algebras, Algebra Universalis 9 (1979), 8-34

Hotz,G.

[1] Schaltkreistheorie, Berlin, New York, 1974

Hu, T.-K.

[1] Stone duality for primal algebra theory, Math. Zeitschr. 110 (1969), 180-198

Imai, Y. and K.Iseki

[1] On axiom systems of propositional calculi, Proc. Japan. Academy, 42 (1966), 19-22

Iseki, K. and S.Tanaka

[1] An introduction to the theory of BCK-algebras, Math. Japonica, 23 (1978), 1-26

Jablonskij, S.W. (Яблонский, С.В.)

[1] Функциональные построения в к-значной логике, Тр.мат. ин-та АН СССР 5I (I958), 5-I42

Jablonskij, S.W., Gawrilov, G.P. and W.B. Kudrjavzev

[1] Boolesche Funktionen und Postsche Klassen, Akademie-Verlag, Berlin 1970

Janov, J.I. (Янов, Ю.И.)

[1] О системах тождеств для алгебр, проблемы кибернетики, 8(I962), 5-I42

Jónsson, B.

[1] Algebras whose congruence lattices are distributive, Math. Scand. 21 (1967), 110-121

Kempe, A.B.

[1] On the relation between the logical theory of classes and the geometrical theory of points, Proc. London Math. Soc. 21 (1890), 147-182

Kiss, E.

[1] Term functions and subalgebras, preprint

Knoebel, R.A.

[1] Functionally complete algebras, Doctoral dissertation, Univ. of California, Berkeley 1965

[2] A class of order preserving algebras which are categorical and precomplete, Notices Amer. Math. Soc. 13 (1966), 226

[3] Representation theorems for varieties generated by single precomplete algebras, Notices Amer. Math. Soc. 23 (1976), A-514

[4] Further conclusions on functional completeness, Fund. Math. 99 (1978), 93-112

[5] Equational classes generated by single functionally precomplete algebras, preprint 1980

Lau, D.

[1] Über Kongruenzen auf maximalen Klassen der k-wertigen Logik, preprint

[2] Bestimmung der Ordnung maximaler Klassen von Funktionen der k-wertigen Logik, Zeitschr. f. Math. Logik und Grundl. der Math. 24 (1978), 79-96

[3] Automorphismen auf den maximalen Klassen der k-wertigen Logik, preprint

[4] Submaximale Klassen von P_3, preprint

Lausch, H. and W. Nöbauer

[1] Algebra of polynomials, North Holland Publ. Comp. New York, 1973

Lugowski, H.

[1] Grundzüge der Universellen Algebra, B.G. Teubner Verlagsgesellschaft, Leipzig 1976

Lukasiewicz, J.
[1] O logice trójwartościowej, Ruch Filozoficzny 5 (1920), 169-170
Lukasiewicz, J. and A.Tarski
[1] Untersuchungen über den Aussagenkalkül, Comptes-rendus des séances de la Société de Sciences et des Lettres de Varsovie, 23 (1930), 30-50
Lyndon, R.C.
[1] Identities in two-valued calculi, Trans. Amer. Math. Soc. 71 (1954), 457-465
[2] Identities in finite algebras, Proc. Amer. Math. Soc. 5 (1954), 8-9
Malcev, A.I. (Мальцев, А.И.)
[1] К общей теории алгебраических систем, матем. сборник, 35,I(1954) 3-20
[2] Итеративные алгебры и многообразия Поста, алгебра и логика 5-2 (1966),5-24
[3] Several remarks on quasivarieties of algebraic systems, Algebra and Logic, 5 (1966), 3-9
[4] Algebraic systems, Akademie-Verlag, Berlin 1973
[5] Итеративные алгебры Поста, Новосибирск 1976
Malcev, I.A. (Мальцев, И.А.)
[1] Конгруенции и автоморфизм на клетках алгебр Поста, алгебра и логика, II,6 (1972), 666-672
Marczewski, E.
[1] Independence in abstract algebras, Colloqu. Math. 14 (1966), 169-188
Martynjuk, V.V. (Мартынюк, В.В.)
[1] Исследование некоторых классов функций в многозначных логиках, Проблемы кибернетики 3 (1960), 49-60
Mc Kenzie, R.
[1] Equational bases and non-modular lattice varieties, Trans. Amer. Math. Soc. 4 (1972)
[2] On minimal, locally finite varieties with permuting congruence relations,preprint
[3] Paraprimal varieties: a study of finite axiomatizability and definable principal congruences in locally finite varieties, Algebra Universalis, 8 (1978), 336-348
Monk, J.D. and F.M. Sioson
[1] On the general theory of m-groups, Trans. Amer. Math. Soc. 14 (1966), 169-188
Mitschke,A.
[1] Implication algebras are 3-permutable and 3-distributive, Algebra Universalis, 1 (1972), 182-186

Murskij, V.L. (Мурский, В.Л.)

[1] Существование в трехзначной локике замкнутого класса с конечным базисом, не имеющих конечной полной системы тождеств, Д.А.Н. СССР, Т.163 (1965), 815-818

[2] Конечная базируемость тождеств и другие свойства почти всех конечных алгебр, Пробл. кибернет. 30 (1975), 43-56

Pigozzi, D.

[1] Minimal, locally finite varieties that are not finitely axiomatizable, Algebra Universalis 9 (1979), 374-390

Pixley, A.F.

[1] Distributivity and permutability of congruence relations in equational classes of algebras, Proc. Amer. Math. Soc. 14(1963), 105-109

[2] Functionally complete algebras generating distributive and permutable classes, Math. Zeitschr. 114 (1970), 361-372

[3] The ternary discriminator function in Universal Algebra, Math. Ann. 191 (1971), 167-180

[4] A note on hemi-primal algebras, Math. Zeitschr. 124 (1972), 213-214

[5] Completeness in arithmetical algebras, Algebra Universalis, 2 (1972), 79-96

[6] Equationally semi-complete varieties, Algebra Universalis, 3 (1974), 323-327

[7] Characterization of arithmetical varieties, preprint

[8] A survey of interpolation in Universal algebra, preprint

Pöschel, R. and L.A.Kalužnin

[1] Funktionen- und Relationenalgebren, Deutscher Verlag der Wissenschaften, Berlin 1978

Post, E.L.

[1] Introduction to a general theory of elementary propositions, Amer. J. Math. 43 (1921), 163-185

[2] Polyadic groups, Trans. Amer. Math. Soc. 48 (1940), 208-350

[3] Two-valued iterative systems of mathematical logics, Ann. Math. Studies 5, Princeton Univ. Press, 1941

Quackenbush, R.W.

[1] Demi-semi-primal algebras and Malcev-type-conditions, Math. Zeitschr. 122 (1971), 166-176

[2] Primality: the influence of Boolean algebras in Universal algebra, Appendix 5 of (Grätzer[1]).

[3] Algebras with minimal spectrum, preprint 1979

[4] A new proof of Rosenberg's primal algebra characterization theorem, preprint 1979

Quackenbush, R.W.
[5] Pseudovarieties of finite algebras isomorphic to bounded distributive lattices, Discrete Mathematics 28 (1979) 189-192

Rasiowa, H.
[1] An algebraic approach to non-classical logics, Warsaw 1974

Rosenberg, I.G.
[1] La structure des fonctions de plusier variables sur ensemble fini, Comptes Rendus de l'Acad.Paris, 260 (1965), 3817-3819
[2] Über die funktionale Vollständigkeit in den mehrwertigen Logiken, Rozpravy Ceskoslovenské Akademie ved, 80 (1970), 3-93
[3] Universal algebras with all operations of bounded range, Coll. Math. 30 (1974), 177-185
[4] Completeness, closed classes and relations in multiple-valued logics, Morgantown 1974, 1-26
[5] Completeness properties of multiple-valued logic algebras, Chapter 6 of Computer Science and Multiple-valued logic, Theory and Applications, D.C.Rine, ed. North Holland, 1977
[6] Functionally complete algebras in congruence distributive varieties, preprint 1979
[7] Functional completeness of single generated or surjective algebras, preprint CRM-856, 1979

Rosser, J.B. and A.R.Turquette
[1] Axiom schemes for m-valued propositional calculi, The Journal of symbolic logic, 10 (1945), 61-82

Rousseau, G.
[1] Completeness in finite algebras with a single operation, Proc. Amer. Math. Soc. 18 (1967), 1009-1013

Salomaa, A.
[1] A theorem concerning the composition of functions of several variables ranging over a finite set, J. Symb. Logic 25 (1960), 203-208

Schweigert, D.
[1] On prepolynomially complete algebras, J.London Math. Soc. 20 (1979), 179-185

Scott, D.
[1] Equationally complete extensions of finite algebras, Nederl. Akad. Wetensch.Proc. Ser.A 59 (1956), 35-38

Senft, J.R.
[1] On weak automorphisms of universal algebras, Dissertationes mathematicae, Warsaw 1970

Sioson, F.M.
[1] Contributions to the theory of primal and functionally complete algebras, Ph.D.thesis, Univ. of California, Berkeley 1960

Smirnov, D.M.(Смирнов, Д.М.)

[1] Многообразия и квазимногообразия алгебр, Colloqu. on Universal algebra, Bolyai,J. Math.Soc., Budapest (1977), 103-104

Slupecki, J.

[1] Completeness criterion for systems of many-valued propositional calculus (Polish), Comptes rendus des Séances de la Societédes Sciences et des Lettres de Varsovie, 32 (1939), 102-109

Swierczkowski, S.

[1] Algebras which are independently generated by every n elements, Fund. Math. 49 (1960), 93-104

Tanaka, S.

[1] On λ-commutative algebras, Math.Seminar notes, 3 (1975), Kobe University

Taylor, W.

[1] Characterizing Malcev conditions, Algebra Universalis, 3 (1973), 351-391

[2] The fine spectrum of a variety, Algebra Universalis, 5 (1975), 263-303

[3] Pure compactifications in quasi-primal varieties, Canad. J. Math. 28 (1976), 50-62

Traczyk, T.

[1] An equational definition of a class of Post algebras, Bull. Acad.Pol.Sc. 12 (1964), 147-149

[2] On the variety of bounded commutative BCK-algebras, preprint 1979

Webb, D.L.

[1] Definition of Post's generalized negative and maximum in terms of one binary operation, Amer. J. Math. 58 (1936), 193-194

Werner, H.

[1] Eine Charakterisierung funktional vollständiger Algebren, Archiv Math. 21 (1970), 381-385

[2] Discriminator algebras, Studien zur Algebra und ihre Anwendungen, Akademie-Verlag, Berlin 1978

Whiteman, A.

[1] Postulates for Boolean algebras in terms of ternary rejection, Bull. Amer. Math. Soc. 43 (1973), 293-298

Yutani, H.

[1] Quasi-commutative BCK-algebras and congruence relations, Math. Seminar notes 5 (1977), 469-480, Kobe University

Zariski, D. and P. Samuel

[1] Commutative algebra, Vol.I, Van Nostrand, Princeton N.J. 1958

Subject Index

algebra
-, affine 83,126
-, arithmetical 33
-, BCK 132
-, Boolean 18
-, bounded commutative BCK 133
-, bounded ordered 87
-, categorical 37
-, commutative BCK 133
-, constantive 31
-, congruence distributive 33
-, congruence permutable 33
-, cryptoprimal 35
-, demiprimal 32
-, demi-semiprimal 32
-, functionally complete 30
-, functionally precomplete 128
-, functionally semicomplete 70
-, functionally incomplete 70
-, hemicategorical 51
-, homogeneous 79
-, implicative BCK 132
-, iterative 75
-, linear hemiprimal 50
-, ordered 87
-, paraprimal 35
-, Post 19
-, plain 39
-, preprimal 33
-, primal 30
-, quasiprimal 32
-, regular subprimal 47
-, rigid 48
-, semicategorical 51
-, semiprimal 32
-, singular subprimal 47
-, splitting 118
-, subprimal 47
-, trivial 30
-, weak homogeneous 79

automorphism
-, weak 75,76
-, clone 75
-, inner 76
centroid element 47
centroid of an algebra 47
clone of operations 30
closed class of functions 26
core 47
equationally maximal 34
equivalence
- of algebras 30
- of categories 105
finite axiomatizability 41
function algebra 75
functionally completeness 26
function
-, affine 83
-, Boolean 11
-, constant 29
-, dual Boolean 53
-, 1-separable Boolean 53
-, flat 98
-, linear 53
-, monotonic 53
-, non-surjective 98
-, self-dual 53
-, sharp 98
-, Sheffer 119
-, surjective 98
-, pre-Sheffer 124
-, pseudo-Sheffer 124
-, wreath 91
fundamental operation 30
gate 15
independent set 61
lattice order 86
Malcev-type-condition 35
maximal function algebra 75
normalizer 77
normal subgroup 76

polynomial 31
- function 30
preprimality 121
primality 119
propositional calculus 11
-, classical 11
-, k-valued 13
-, implicative 132
projections 26
pseudovariety 118
quasivariety 118
relation
-, bounded order 86
-, central 66
-, h-elementary 66
-, h-universal 66
-, primitive h-elementary 66
-, totally reflexive 29
-, totally symmetric 66
skew congruences 39
subalgebra
-, rectangular 96
-, reflexive 92
-, symmetric 92
-, transitive 92
superposition of functions 26
spectrum of a variety 60
--, fine 59
--, minimal 61
switching circuit 15
term 31
- function 30
ternary discriminator 42
variety 34
-, arithmetical 36
-, congruence permutable 34
-, congruence distributive 34
-, degenerate 111
-, n-distributive 35
-, semidegenerate 111

MATHEMATICAL RESEARCH - MATHEMATISCHE FORSCHUNG

Band 1

INVERSE AND IMPROPERLY POSED PROBLEMS IN DIFFERENTIAL EQUATIONS

Proceedings of the Conference on Mathematical and Numerical Methods, held in Halle/Saale (GDR) from May 29 to June 2, 1979
Edited by Gottfried Anger
1979. 296 Seiten - 18 Abb. - 5 Tab. - gr. 8° - 37,-- M
Bestell-Nr. 762 744 1 (2182/1) in englischer Sprache

Band 2

FUNDAMENTALS OF COMPUTATION THEORY FCT '79

Proceedings of the Conference on Algebraic, Arithmetic and Categorial Methods in Computation Theory, held in Berlin/Wendisch-Rietz (GDR) September 17-21, 1979
Edited by Lothar Budach
1979. 576 Seiten - 27 Abb. - 2 Tab. - gr. 8° - 72,-- M
Bestell-Nr. 762 806 5 (2182/2) in englischer Sprache

Band 3

DECIDABILITY AND GENERALIZED QUANTIFIERS

By Andreas Baudisch, Detlef Seese, Hans-Peter Tuschik, and Martin Weese
1980. XII, 235 Seiten - 16 Abb. - gr. 8° - 31,-- M
Bestell-Nr. 762 808 1 (2182/3) in englischer Sprache

Band 4

ZUR ELASTIZITÄTS- UND THERMOELASTIZITÄTSTHEORIE

Potentialtheoretische Methoden in Anwendung auf verschiedene Modelle
Von Lothar Jentsch und Johannes Maul
1980. 157 Seiten - gr. 8° - 20,-- M
Bestell-Nr. 762 810 2 (2182/4)

Band 5

SYSTEMS ANALYSIS AND SIMULATION 1980

Proceedings of the International Symposium held at Berlin (GDR) September 1-5, 1980
Edited by Achim Sydow
1980. 496 Seiten - 111 Abb. - 11 Tab. - gr. 8° - 62,-- M;
Ausland 96,-- M - Bestell-Nr. 762 811 0 (2182/5) in englischer Sprache

Band 6

NICHTLINEARE DISKRETE OPTIMIERUNG

Von Eberhard Girlich und M. M. Kowaljow
1981. 218 Seiten - gr. 8° - 28,-- M - Bestell-Nr. 762 925 2 (2182/6)

Band 7

COMPLETENESS PROBLEMS IN THE STRUCTURAL THEORY OF AUTOMATA

By Jürgen Dassow
1981. 148 Seiten - gr. 8° - 20,-- M - Bestell-Nr. 762 951 9 (2182/7) in englischer Sprache

Band 8

DARSTELLUNG VON LÖSUNGEN LINEARER ELLIPTISCHER DIFFERENTIALGLEICHUNGEN

Von Günther Wildenhain
1981. 92 Seiten - gr. 8° - 12,-- M - Bestell-Nr. 763 003 4 (2182/8)

Band 9

INTRODUCTION TO THE SPECTRAL THEORY OF OPERATORS IN SPACES WITH AN INDEFINITE METRIC

By I. S. Iohvidov, M. G. Krein, and H. Langer
1982. 120 Seiten - gr. 8° - 15,-- M - Bestell-Nr. 763 041 3 (2182/9)

This paper offers an introduction to the theory of the simplest class of spaces with indefinite scalar product, the spaces $\Pi_\varkappa$ or Pontrjagin spaces, and the linear operators acting in these spaces. It was written on the basis of an article by the first two authors which appeared in 1956 and in English translation in 1960.

Compared with this article, the present paper has been enlarged with many respects. In particular, it contains a detailed study of various new classes of operators in spaces $\Pi_\varkappa$, e.g. of expansive, contractive and dissipative operators. The central results are about the spectral properties and special invariant subspaces of these operators. Results of this type have been found important applications in various branches of mathematical analysis during the last 20 years.

Band 10

NONLINEAR TIME-DISCRETE SYSTEMS

A General Approach by Nonlinear Proposition
By Michael Gössel
1982. 116 Seiten - gr. 8° - 15,-- M - Bestell-Nr. 763 076 3 (2182/10)

In this book a nonlinear systems theory as a theory of nonlinear superposition is proposed.
Linear systems theory is essentially based on linear superposition. Therefore it seems to be natural to generalize this theory by generalizing the superposition principle as it was done in this book for time-discrete systems or automata.
The superposition of automata with respect to an arbitrary pair of operations (without memory) and a pair of automata (with memory), respectively, is investigated. The concept of invariant relations for automata is also discussed.
Different classes of nonlinear systems like binary automata, generalized linear automata, invertible automata are considered in detail.
It is shown how to extend the description of the input-output-behaviour by impulse responses, well-known in linear systems theory, to nonlinear systems.
Homomorphic systems, useful in seismology and nonlinear superponable differential equations are treated in an appendix.

Im AKADEMIE-VERLAG sind in der Reihe
MATHEMATISCHE LEHRBÜCHER UND MONOGRAPHIEN
in den letzten Jahren erschienen:

Abteilung I: Mathematische Lehrbücher

Band IV
N. I. Achieser, I. M. Glasmann: Theorie der linearen Operatoren im Hilbert-Raum (Übersetzung aus dem Russischen, herausgegeben von H. Baumgärtel), 8., neubearbeitete und erweiterte Auflage, 1981

Band VI
I. P. Natanson: Theorie der Funktionen einer reellen Veränderlichen (Übersetzung aus dem Russischen, herausgegeben von K. Bögel), 6. Auflage, 1978

Band VIII
L. A. Ljusternik, W. I. Sobolew: Elemente der Funktionalanalysis (Übersetzung aus dem Russischen, herausgegeben von K. Gröger), 6. Auflage, 1979

Band 27
A. A. Borowkow: Wahrscheinlichkeitstheorie. Eine Einführung (Übersetzung aus dem Russischen, herausgegeben von P. Franken), 1975

Band 28
J. A. Rosanow: Stochastische Prozesse. Eine Einführung (Übersetzung aus dem Russischen, herausgegeben von U. Küchler), 1976

Band 29
M. Petrich: Lectures in Semigroups, 1977

Band 30
S. G. Michlin: Partielle Differentialgleichungen in der mathematischen Physik (herausgegeben von S. Prößdorf), 1978

Band 31
S. W. Jablonski, O. B. Lupanow (Hrsg.): Diskrete Mathematik und mathematische Fragen der Kybernetik (Übersetzung aus dem Russischen, herausgegeben von G. Burosch und H. Kiesewetter), 1980

Band 32
G. P. Klimow: Bedienungsprozesse (herausgegeben von D. König), 1978

Band 33
G. Eisenreich: Lineare Algebra und analytische Geometrie, 1980

Band 34
G. P. Tolstow: Maß und Integral (Übersetzung aus dem Russischen, herausgegeben von R. Kühne), 1981

Abteilung II: Mathematische Monographien

Band 43/44/45
K. M. S. Humak: Statistische Methoden der Modellbildung I, II, III, 1977, 1982, 1983

Band 46
B. W. Schulze, G. Wildenhain: Methoden der Potentialtheorie für elliptische Differentialgleichungen beliebiger Ordnung, 1977

Band 47/48
H. Bandemer u. a.: Theorie und Anwendung der optimalen Versuchsplanung. I — Theorie, 1978. II — Anwendungen, 1980

Band 49
R. E. Barlow, B. Proschan: Statistische Theorie der Zuverlässigkeit (Übersetzung aus dem Englischen, herausgegeben von P. Franken), 1980

Band 50
A. D. Wentzell: Theorie zufälliger Prozesse (Übersetzung aus dem Russischen, herausgegeben von J. Engelbert und J. Groh), 1979

Band 51
K. Lommatzsch (Hrsg.): Anwendungen der linearen parametrischen Optimierung, 1979

Band 52
S. G. Michlin, S. Prößdorf: Singuläre Integraloperatoren, 1980

Band 53
P. Franken, D. König, U. Arndt, V. Schmidt: Queues and Point Processes, 1981

Band 54
J. W. Gelman, W. G. Mazja: Abschätzungen für Differentialoperatoren im Halbraum (herausgegeben von G. Wildenhain), 1981

Band 55
S. Rempel, B.-W. Schulze: Index Theory of Elliptic Boundary Problems, 1982

Band 56/57
B. W. Gnedenko, D. König (Hrsg.): Handbuch der Bedienungstheorie I, II, 1982, 1983

Band 58
B. Bank, J. Guddat, D. Klatte, B. Kummer, K. Tammer: Nonlinear Parametric Optimization, 1982

Band 59
H. Baumgärtel, M. Wollenberg: Mathematical Scattering Theory, 1982

In the theory of two- and multiple-valued logics, in universal algebra and in the structural theory of automata the following functional completeness problem plays an essential role: whether or not the set of all on a finite set defined functions can be generated by superposition from a given set of functions defined on this set. The development of an algebraic structure theory for preprimal algebras appears as a problem of growing importance in connection with an algebraic approach to the functional completeness problem since the set of all term functions of a preprimal algebra is a maximal class of functions. By determination of all maximal classes E. L. Post, S. V. Jablonskij and I. G. Rosenberg solved the completeness problem in two- and multiple-valued logics. Completeness questions in universal algebras were first investigated by A. L. Foster and A. F. Pixley. The aim of this monograph is an extensive description of the actual state of the research into preprimal algebras on a unified base covering both dispositions.

In der Theorie der zwei- und mehrwertigen Logiken, in der universellen Algebra und in der strukturellen Automatentheorie spielt das folgende Problem der funktionalen Vollständigkeit eine bedeutende Rolle: Kann die Menge aller auf einer endlichen Menge definierten Funktionen aus einer ihrer Teilmengen durch Superposition erzeugt werden oder nicht. Im Zusammenhang mit einem algebraischen Zugang zum Problem der funktionalen Vollständigkeit ist die Entwicklung einer algebraischen Strukturtheorie präprimaler Algebren von wachsender Bedeutung, da die Menge aller Termfunktionen einer präprimalen Algebra eine maximale Klasse von Funktionen ist. Mit Hilfe der maximalen Klassen gelang es Post, Jablonskij und Rosenberg, das Vollständigkeitsproblem in den zwei- und mehrwertigen Logiken zu lösen. Vollständigkeitsfragen in universellen Algebren wurden zuerst von A. L. Foster und A. F. Pixley untersucht. Das Anliegen dieser Monographie besteht darin, den gegenwärtigen Stand der Untersuchung präprimaler Algebren auf einer beide Ansätze vereinigenden Grundlage darzustellen.

www.ingramcontent.com/pod-product-compliance
Lightning Source LLC
LaVergne TN
LVHW060621300926
849162LV00008B/1680